Springer Tracts in Modern Physics
Volume 156

Managing Editor: G. Höhler, Karlsruhe

Editors: J. Kühn, Karlsruhe
Th. Müller, Karlsruhe
R. D. Peccei, Los Angeles
F. Steiner, Ulm
J. Trümper, Garching
P. Wölfle, Karlsruhe

Honorary Editor: E. A. Niekisch, Jülich

Springer
*Berlin
Heidelberg
New York
Barcelona
Hong Kong
London
Milan
Paris
Singapore
Tokyo*

Springer Tracts in Modern Physics

Springer Tracts in Modern Physics provides comprehensive and critical reviews of topics of current interest in physics. The following fields are emphasized: elementary particle physics, solid-state physics, complex systems, and fundamental astrophysics.
Suitable reviews of other fields can also be accepted. The editors encourage prospective authors to correspond with them in advance of submitting an article. For reviews of topics belonging to the above mentioned fields, they should address the responsible editor, otherwise the managing editor. See also http://www.springer.de/phys/books/stmp.html

Managing Editor

Gerhard Höhler

Institut für Theoretische Teilchenphysik
Universität Karlsruhe
Postfach 69 80
D-76128 Karlsruhe, Germany
Phone: +49 (7 21) 6 08 33 75
Fax: +49 (7 21) 37 07 26
Email: gerhard.hoehler@physik.uni-karlsruhe.de
http://www-ttp.physik.uni-karlsruhe.de/

Elementary Particle Physics, Editors

Johann H. Kühn

Institut für Theoretische Teilchenphysik
Universität Karlsruhe
Postfach 69 80
D-76128 Karlsruhe, Germany
Phone: +49 (7 21) 6 08 33 72
Fax: +49 (7 21) 37 07 26
Email: johann.kuehn@physik.uni-karlsruhe.de
http://www-ttp.physik.uni-karlsruhe.de/~jk

Thomas Müller

Institut für Experimentelle Kernphysik
Fakultät für Physik
Universität Karlsruhe
Postfach 69 80
D-76128 Karlsruhe, Germany
Phone: +49 (7 21) 6 08 35 24
Fax: +49 (7 21) 6 07 26 21
Email: thomas.muller@physik.uni-karlsruhe.de
http://www-ekp.physik.uni-karlsruhe.de

Roberto Peccei

Department of Physics
University of California, Los Angeles
405 Hilgard Avenue
Los Angeles, CA 90024-1547, USA
Phone: +1 310 825 1042
Fax: +1 310 825 9368
Email: peccei@physics.ucla.edu
http://www.physics.ucla.edu/faculty/ladder/peccei.html

Solid-State Physics, Editor

Peter Wölfle

Institut für Theorie der Kondensierten Materie
Universität Karlsruhe
Postfach 69 80
D-76128 Karlsruhe, Germany
Phone: +49 (7 21) 6 08 35 90
Fax: +49 (7 21) 69 81 50
Email: woelfle@tkm.physik.uni-karlsruhe.de
http://www-tkm.physik.uni-karlsruhe.de

Complex Systems, Editor

Frank Steiner

Abteilung Theoretische Physik
Universität Ulm
Albert-Einstein-Allee 11
D-89069 Ulm, Germany
Phone: +49 (7 31) 5 02 29 10
Fax: +49 (7 31) 5 02 29 24
Email: steiner@physik.uni-ulm.de
http://www.physik.uni-ulm.de/theo/theophys.html

Fundamental Astrophysics, Editor

Joachim Trümper

Max-Planck-Institut für Extraterrestrische Physik
Postfach 16 03
D-85740 Garching, Germany
Phone: +49 (89) 32 99 35 59
Fax: +49 (89) 32 99 35 69
Email: jtrumper@mpe-garching.mpg.de
http://www.mpe-garching.mpg.de/index.html

Nikolai N. Ledentsov

Growth Processes and Surface Phase Equilibria in Molecular Beam Epitaxy

With 17 Figures

 Springer

Professor Nikolai N. Ledentsov
Technische Universität Berlin
Institut für Festkörperphysik
Hardenbergstrasse 36
D-10623 Berlin, Germany
Email: leden@sol.physik.TU-Berlin.de
and
Abraham F. Ioffe Physical-Technical Institute
Politekhnicheskaya 26
194021 St. Petersburg, Russia

Physics and Astronomy Classification Scheme (PACS):
81.15.Hi, 81.10.Aj, 81.05.Ea, 81.05.Dz, 82.60.-s, 68.10.Jy, 68.35.Fx, 66.30, 68.35.ct

ISSN 0081-3869
ISBN 3-540-65794-0 Springer-Verlag Berlin Heidelberg New York

Library of Congress Cataloging-in-Publication Data.
Ledentsov, Nikolai N. Growth processes and surface phase equilibria in molecular beam epitaxy/Nikolai N. Ledentsov. p. cm. – (Springer tracts in modern physics, ISSN 0081-3869; v. 156). Includes bibliographical references and index. ISBN 3-540-65794-0 (hc.: alk. paper). 1. Molecular beam epitaxy. 2. Crystal growth. 3. Semiconductor doping. I. Title. II. Series: Springer tracts in modern physics; 156.   QC1.S797 vol. 156   [QC611.6.M64]   539 s–dc21   [548.5'175]   99-30928

This work is subject to copyright. All rights are reserved, whether the whole or part of the material is concerned, specifically the rights of translation, reprinting, reuse of illustrations, recitation, broadcasting, reproduction on microfilm or in any other way, and storage in data banks. Duplication of this publication or parts thereof is permitted only under the provisions of the German Copyright Law of September 9, 1965, in its current version, and permission for use must always be obtained from Springer-Verlag. Violations are liable for prosecution under the German Copyright Law.

© Springer-Verlag Berlin Heidelberg 1999
Printed in Germany

The use of general descriptive names, registered names, trademarks, etc. in this publication does not imply, even in the absence of a specific statement, that such names are exempt from the relevant protective laws and regulations and therefore free for general use.

Typesetting: Data conversion by Springer-Verlag, Heidelberg
Cover design: *design & production* GmbH, Heidelberg
Computer-to-plate and printing: Mercedesdruck, Berlin
Binding: Universitätsdruckerei H. Stürtz AG, Würzburg

SPIN: 10698326   56/3144/tr - 5 4 3 2 1 0 – Printed on acid-free paper

Preface

There exists at present a growing interest in the better understanding of growth mechanisms at crystal surfaces. There are two main reasons for this: first, the range of materials used for epitaxial growth has been significantly expanded in recent years; and, second, numerous new growth-related phenomena have been discovered in the last few years that have captured the attention of researchers. This is particularly true for the rapidly expanding field of self-organized growth, which allows direct fabrication of heterostructures with reduced dimensionality: quantum wires and quantum dots.

A significant proportion of the latest developments, both in expanding the field of materials and in exploiting new ways of nanostructure fabrication, is related to molecular beam epitaxy (MBE). This technique is considered to be particularly promising, because it allows in a natural way numerous analytical facilities for precise control over surface morphology and growth parameters.

In contrast, there also exists a significant controversy in understanding the major growth processes in MBE. In this book we shall summarize the results of the thermodynamic model for the most important MBE growth-related processes and their interconnection. Deposition, thermal etching, doping, segregation and diffusion of both main elements and impurities are each addressed within the framework of a unified thermodynamic model, as are the influence of strain in the growing film on the system parameters and some other observable effects.

The book contains summaries of experimental results and theoretical calculations for some of the key semiconductor materials systems, and it can be used as a practical guide for crystal growers. It should also be useful for students, engineers and researchers interested in a better understanding of modern growth techniques and growth mechanisms, which are crucial for practical fabrication and the understanding of the properties of the structures used for device applications and fundamental research.

The author is grateful to S. V. Ivanov and P. S. Kop'ev for numerous useful discussions and to R. Heckingbottom for encouraging support at the early stage of this work. Helpful discussions with Zh. I. Alferov, D. Bimberg, A. Yu. Egorov, K. R. Evans, F. Heinrichsdorff, A. R. Kovsh, V. M. Ustinov and A. E. Zhukov are gratefully acknowledged.

My colleagues at TU Berlin and Abraham Ioffe Institute and I are thankful to the Russian Foundation of Basic Research, International Science Foundation, INTAS, NATO and Volkswagen Foundation for their support of our work. The author is personally grateful to the Alexander von Humboldt Foundation and the Guest Professorship Program of DAAD.

Berlin and St. Petersburg
April 1999

Nikolai N. Ledentsov

Contents

1. Introduction 1
2. Basics of MBE Growth 3
   2.1 MBE Apparatus 3
   2.2 Understanding of MBE Growth Processes 5
   2.3 Solid–Vapor Equilibrium for Binary Compounds 8
   2.4 Liquid–Solid–Vapor Equilibrium for Binary Compounds 15
   2.5 Particular III–V Materials 16
      2.5.1 AlAs 16
      2.5.2 InAs 17
      2.5.3 InP 18
      2.5.4 GaP 19
   2.6 Solid–Vapor Equilibrium for Ternary Compounds 20
   2.7 Liquid–Solid–Vapor Equilibrium for Ternary Compounds: Surface Segregation of More Volatile Elements 22

3. Doping and Impurity Segregation Effects in MBE 33
   3.1 Point-Defect Equilibria in MBE 33
   3.2 Impurity Incorporation in MBE 36
      3.2.1 General Consideration 36
      3.2.2 Manganese Doping of GaAs 37
      3.2.3 GaAs Doping with Zn, Cd, Pb, Mg 39
      3.2.4 GaAs Doping with S, Se, Te 40
      3.2.5 GaAs Doping with Amphoteric Impurities: Ge, Si, Sn . 41
   3.3 Impurity Segregation in MBE 43
   3.4 Interplay Between Impurity Segregation and Diffusion in MBE 47

4. Influence of Strain in the Epitaxial Film on Surface-Phase Equilibria 59
   4.1 MBE Growth of Lattice-Mismatched Binary Compound 59
   4.2 Growth of Lattice-Matched Solid Solution Formed From Lattice-Mismatched Binaries 66

5. II–VI Materials 71
6. Conclusion 75
Index 83

1. Introduction

The question of the relative role of kinetics and thermodynamics in modern epitaxial growth techniques is of principal importance, as it defines the primary alternative ways of composition, point-defect concentration and growth-rate control. It plays an important role in optimizing growth parameters so as to obtain high-quality structures with planar or periodically modulated interfaces.

In the past there has been a lot of controversy concerning the possibility of applying thermodynamics to growth and doping in molecular beam epitaxy (MBE) – see, e.g. [39] and references therein. Despite the discussions concerning the relative importance of kinetics and thermodynamics being far from over even today, in the 1980s MBE was developing experimentally very successfully and most of the problems that existed in the early stages of this technology were solved empirically. Further success in device applications, particularly for heterostructures with ultrathin layers in the GaAs–(Al,Ga)As and (In,Ga,Al)As materials systems, gave an impression that most of the problems for this technology were either solved or were shortly to be solved without having to delve too deeply into the details of growth fundamentals. The dispute concerning the relative importance of thermodynamics and kinetics in MBE was then considered as being solely academic in nature.

This situation has changed dramatically in recent years, when, first of all, the range of materials extensively used in MBE growth was dramatically expanded, and an explosion of interest in the spontaneous formation of ordered nanostructures on crystal surfaces occurred. Approaches to describe self-organization of nanostructures on crystal surfaces can be also roughly divided in two groups: kinetic models [5, 6, 11, 72] and thermodynamic models [68, 69, 75, 76, 99, 106–110, 120]. A brief review on these models can be found in a recent reference book on quantum dots [9].

Generally, the thermodynamic approaches are based on an assumption that all kinetic processes are fast enough and that the system can reach its equilibrium (lowest energy) state. Kinetic approaches, on the other hand, assume that the equilibrium state is never reached and that all the main properties of the system (e.g. concentrations and types of point defects, and the geometrical size and shape of spontaneously formed nanoislands) are defined by particular kinetic pathways. This divergence of approach is to some degree false, as any realistic kinetic model should lead to the same result as a thermodynamic model does, when the growth rate is extrapolated to zero. Thus, in the general case, the knowledge of the thermodynamically favorable state is important to construct a proper kinetic pathway. The recent interest in the self-organized growth of nanostructures has revived interest in the interconnection between kinetics and thermodynamics in MBE still further.

Here we give a review of a unified thermodynamic model of major growth-related effects in MBE, such as the condensation or evaporation of main elements and impurities, surface segregation of more volatile main elements, and the segregation of impurities.

2. Basics of MBE Growth

2.1 MBE Apparatus

First of all we shall address ourselves to the basics of molecular beam epitaxy. MBE is a technique for the epitaxial growth of materials by means of the chemical interaction of one or several molecular or atomic beams of different intensities and compositions that occurs on the surface of a heated single-crystalline substrate.

A schematic representation of suitable MBE apparatus is shown in Fig. 2.1. The source materials are placed in evaporation cells that are composed of a crucible of shape and dimensions ensuring the required angular distribution of atoms or molecules in a beam, a resistive heater, and thermal screens. The angular distribution of the beam and the distances between the sources and the substrate determine the homogeneity of the parameters of epilayers and heterostructures grown by this technique. A manipulator with

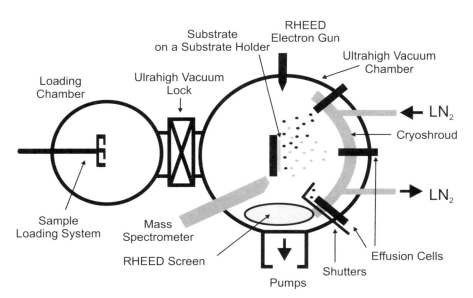

Fig. 2.1. Schematic view of MBE apparatus

a substrate holder is used to enable the required position of the substrate relative to the cells to be obtained, and to heat it to the necessary temperature. The homogeneity of the grown films is often improved by rotation of the substrate.

The molecular beam condition (where the free path of the particle is larger than the geometrical size of the chamber) is ensured, for typical distances between the sources and the substrate, if the total pressure does not exceed 10^{-4} Torr. However, all MBE systems are, as a matter of course, provided with a means of reaching and maintaining an ultrahigh vacuum ($\sim 10^{-11}$ Torr) and the operation is usually oil-free. One of the reasons why MBE systems have to be oil-free is the need to ensure that the substrate is atomically clean before growth. Even so, a low level of background doping and control over the properties of the grown materials and structures can be assured only if uncontrolled fluxes of atoms reaching the substrate surface are as weak as possible. An ultrahigh vacuum is essential for this purpose, but it is not a sufficient condition; two other conditions need to be applied. First, any vacuum represents an equilibrium between the rate of gas evolution and the rate of pumping, so that it is necessary to use construction and crucible materials with the lowest rate of gas evolution. The usual crucible material for MBE of III–V compounds is boron nitride, which combines a low rate of gas evolution with weak chemical activity right up to temperatures of the order of 1500 °C. Second, it is important to ensure cryogenic screening around the substrate so as to minimize stray fluxes of atoms and molecules from the walls of the chamber, which are at room temperature, and from the heated components of the apparatus. It is necessary also to monitor not only the total pressure but also the partial composition of the atmosphere and, if necessary, to alter it (by, for example, exposure of the substrate surface to hydrogen flux) in order to reduce the partial pressures of the most active components of the background atmosphere and so hinder incorporation of the undesirable impurities as a result of the reactions on the substrate surface. Only ultrapure materials can be used as source materials.

One advantage of the MBE technology that has made it very popular among crystal growers and device engineers is the intrinsic feasibility of controlling the profile of the composition and doping of a growing structure at monolayer level. The molecular beam regime during growth, which excludes any interaction between molecular beams, in combination with the relatively low growth rate, ensure this feasibility. An abrupt change in the composition and/or the degree and nature of doping are achieved by opening or closing the relevant fluxes by the shutters with which each cell is supplied. The operation time of a shutter (< 0.1 s) is usually considerably less that the time needed to grow one monolayer (typically 1–5 s). Variation of the temperatures of the cells and, consequently, of the intensities of molecular fluxes and the corresponding variation (if necessary) of the growth rate potentially provides an opportunity of establishing any specified profile of composition and doping in the film.

Ultrahigh-vacuum conditions and the open growth surface provide extensive opportunities for controlling the technological process at all its stages. Preliminary preparation of an atomically clean defect-free substrate surface is exceptionally important in this technology. This process normally includes (where substrates are used that are not epi-ready) chemical–mechanical treatment of the substrate in a polishing etchant, passivation of the surface by oxidation, and removal of a protective oxide film in a vacuum chamber during heating.

The MBE chamber is equipped with a reflection high-energy electron diffraction (RHEED) system and mass spectrometers for monitoring the beams, their molecular composition, and the residual atmosphere. The chamber also contains ionisation gauges for monitoring the fluxes. A modification of the RHEED technique involving a study of oscillations of the intensity of diffraction reflections during growth makes it possible to monitor not only the reconstruction of the film surface but also its smoothness at a monolayer level, the surface diffusion length of migrating atoms, and the deposition rate.

2.2 Understanding of MBE Growth Processes

Early MBE investigations into the conventional III–V material system revealed that, at the relatively low substrate temperatures typically used during this period, all atoms of a group-III element have unity sticking coefficient to the substrate surface [15]. It was also found that group-V atoms (or molecules) do not stick to the surface in the absence of a flux of group-III atoms. Therefore, the growth rate is governed completely by the rate of arrival of group-III atoms on the substrate surface, and the excess group-V atoms are desorbed from the surface. In contrast with group-III elements arriving on the surface in the form of atoms, group-V elements may reach the surface in the form of various molecules. For example, in the case of the growth of GaAs by MBE the arsenic may arrive on the surface in the form of tetramers As_4 (when the beam source is heated metallic arsenic), in the form of dimeric molecules As_2 (when the source is crystalline GaAs, or when a high-temperature cracker is used for dissociation of tetrameric molecules), or in the form of As atoms (when the source is a high-temperature device for dissociating arsine AsH_3). The arsenic molecules reaching the surface become adsorbed, participate in association–dissociation reactions on the surface, interact with Ga atoms to form GaAs, or are desorbed from the surface in the form of one or another molecule.

The details of these processes are actually not very well understood at present. When the total flux of arsenic atoms reaching the surface is less than the flux of gallium atoms, droplets of liquid gallium are formed on the surface. Early experiments led to the conclusion that the minimum flux of the As_2 molecules (J_{As_2}) necessary to maintain growth is $1/2\, J_{Ga}$, i.e. that each As atom from an As_2 molecule is used to form GaAs. In contrast, studies of beams of As_4 molecules show that at best only half the As atoms interact with gallium to form GaAs and, therefore, the As_2 molecules are more effective in the growth of GaAs by the MBE method [28]. Since then, other more accurate measurements have shown that the effectiveness of As_2 and As_4 molecules in MBE of InAs, GaAs and (In,Ga,Al)As is the same for each type of molecule [103] and it exceeds a ratio of 55% [138]. It has also been found that the ratio of the various arsenic species in a flux from the surface of GaAs in the typical MBE range is governed by the substrate temperature and is independent of the type of arsenic molecule incident on the surface. At substrate temperatures T_S below 300 °C there is a predominance of As_4 molecules, whereas at $T_S \geq 400\,°C$ the bulk is predominantly As_2 molecules [28].

During the first stage of growth studies, it was generally assumed that MBE is a process that shows fundamental nonequilibrium and that thermodynamic approaches are completely inappropriate, so that information on the growth processes can be obtained only by investigation of the kinetics of the specific reactions on the surface. Experiments using modulated molecular beams led to a theory according to which the main role in MBE of III–V compounds is played by elementary adsorption, migration and desorption processes of atoms and molecules [28]. However, in the case of binary compounds the real growth picture is much more complex and cannot be analyzed by simple representations [74]. For example, the probability of jumps of gallium atoms between the crystal lattice sites on the surface depends on the number and configuration of bonds with arsenic atoms at each specific site.

The existing kinetic models of MBE of GaAs usually ignore the processes of detachments of atoms from islands and surface steps, the thermal generation of surface vacancies, any concentration of arsenic adatoms on the surface, the desorption of gallium atoms, processes associated with adsorption, segregation and desorption of impurities, possible reactions between impurities and the main elements, the transformation of neutral atoms of impurities into ions in the course of their incorporation in the crystal lattice, elastic interactions between steps and facets, etc. Even simplified models require very complex calculations [19]. It should also be borne in mind that changes in any one of the parameters of the system (such as the flux of a group-V element) may alter the other parameters (such as the concentrations of adatoms and surface vacancies, the densities of growth steps), so that the transient characteristics observed in experiments with modulated molecular beams are difficult to interpret.

Therefore, in spite of the obvious importance of the kinetic models in particular cases for improving our understanding of the growth processes at microscopic level during MBE, these models have so far been unsuccessful in predicting quantitative dependencies. For example, such models provide only a basic qualitative description of changes in the rates of growth or evaporation of III–V compounds on T_S and on the intensity of the flux of group-V molecules. Considerable difficulties are encountered in attempts to describe the dependence of the composition of a multicomponent solid solution on the growth parameters, particularly in the case of compounds containing two group-V elements. These kinetic models contain a large number of parameters that can usually only be estimated, so that the authors proposing these models frequently arrive at opposite conclusions, and on the whole the kinetic models of composition and growth rate control are well behind the experimental results available. In view of this situation, there have been many attempts to develop a complete thermodynamic description of MBE [35–37, 39, 50, 52, 59, 62, 63, 104, 111].

The fact that the MBE process occurs under strongly nonequilibrium conditions for group-III elements does not provide formal justification for the conclusion that a thermodynamic approach cannot be used to describe an MBE on the basis of equations of mass action in combination with equations describing the conservation of the mass of interacting elements. This approach is in fact used widely – and successfully so – in the description of many chemical reactions that occur under strongly nonequilibrium conditions. In the case of sublimation of binary compounds (which is again a process that strongly demonstrates nonequilibrium), the validity of the thermodynamic approach has been confirmed by numerous experimental data well before the appearance of MBE (see, for example, [65]). This has provided a stimulus for attempts to develop a thermodynamic description of the MBE processes and to utilize the extensive experimental and theoretical data already accumulated in vapor (VPE) and liquid phase (LPE) epitaxy.

At first sight, in the case of MBE a system cannot be described by thermodynamic representations because its different parts are at different temperatures. However, if we assume that the thermalization times of atoms and molecules reaching the substrate surface in the system are considerably less than the time required to grow one monolayer, then the temperature of the system can be assumed to be the temperature of the substrate. The validity of this assumption is confirmed by the fact that the fluxes of atoms and molecules from the substrate have its temperature irrespective of the temperature of the fluxes arriving at the surface; and the nature of the arsenic molecules in the flux coming from the substrate is independent on the nature of the arsenic molecules reaching the surface [28].

Comparable efficiencies of different arsenic molecules used in MBE are also pointing to a quasi-equilibrium nature of the growth process. Moreover, it has also been shown that the best-quality structures – nearly ideal quantum wells (QWs), modulation-doped structures with ultrahigh electron mobility ($10^7 \text{cm}^2\text{V}^{-1}\text{s}^{-1}$) [29], single-QW heterostructure lasers with threshold current density as low as $43 \, \text{A/cm}^2$ [4] – were obtained by using As_4 beams.

High-quality GaAs layers have been reported to be grown by using As_4 beams at a temperature as low as 200 °C [40], which means that the interaction of As_4 molecules with a GaAs substrate is not hindered significantly even at this temperature. So one can assume that the kinetic limitations for arsenic tetramer decomposition are not very significant, at least in the temperature range of interest.

We consider an equilibrium state in a system comprising the gaseous phase and the surrounding volume, the temperature of which is governed by the substrate temperature. The equilibrium partial pressures are those pressures that represent the fluxes of atoms and molecules leaving the substrate surface. A similar approach has been used in a thermodynamic description of the process of the sublimation of binary compounds [65]. In view of the principal importance of the thermodynamic approach, we shall consider the most important aspects of the thermodynamic description and their interconnection with the main growth-related phenomena.

The key points that have to be dealt with in any thermodynamic model for MBE are:

- how MBE growth conditions relate to those in VPE and LPE;
- which parameters of the growth process define stoichiometry of the growing compound;
- how this stoichiometry is related to the impurity incorporation processes;
- what are the optimal growth regimes for particular compounds; and
- in the case of the growth of multicomponent solid solutions, what the parameters are that govern the composition.

The most important question in the development of any model is associated with the basic growth parameters, which derive the final state of the system. In some kinetic models of GaAs growth for MBE, these are: the substrate temperature; the flux ratio of impinging As and Ga atoms; and the growth rate. In the thermodynamic model described in [36], the primary parameters are the substrate temperature and the As–Ga flux ratio. In the thermodynamic model presented in [62], the parameters are the substrate temperature and the effective As pressure. The meaning of the last parameter will be clear from the following description.

2.3 Solid–Vapor Equilibrium for Binary Compounds

Let us consider the T–x projection and the P–x cross-section of the phase diagram of GaAs, schematically represented in Fig. 2.2. In the T–x projection (Fig. 2.2a), the shaded area represents GaAs(s) that is in equilibrium with the gas phase of particular composition (see also Fig. 2.2b). The borders of this area correspond to the points of transition to a three-phase region where, in addition to GaAs, precipitates of the other phase appear. On the left-hand side of the diagrams these are Ga precipitates, and on the right-hand side they are arsenic precipitates. Within the shaded area the deviation from the exact $Ga_{0.5}As_{0.5}$ stoichiometry is realized via formation of intrinsic point defects.

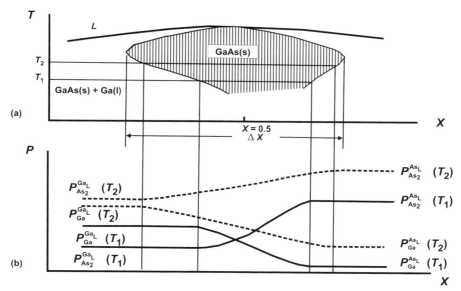

Fig. 2.2. Schematic representation of (a) the T–x projection and (b) the P–x cross-section of the GaAs phase diagram

According to the phase rule, the sum of the number of phases (ϕ) and the number of degrees of freedom (ν) in the system equals the number of components (K) plus the number of parameters that define the state of the system. In the case where the state of the system can be changed only by temperature (T) and pressure (P), and the volumes of the phases are large enough to neglect the surface of interface energy, we can write:

$$\nu = K + 2 - \phi. \tag{2.1}$$

Inside the shaded area of Fig. 2.2a the solid phase GaAs(s) is in equilibrium with the gas phase ($K = 2, \phi = 2$), and thus from (2.1) $\nu = 2$, i.e. the state of the system can be changed by independent adjustment of two parameters – in this case pressure and temperature.

At the boundary of the shaded area of Fig. 2.2a (the solidus curve), there are three phases, which are each in equilibrium. Thus $\nu = 1$, and the system is completely defined by just one parameter (P or T), which are strictly related:

$$P = f(T). \tag{2.2}$$

Let us consider first the P–x cross-sections of the phase diagram at two different temperatures T_1 and T_2. We shall consider the reactions between the main components in MBE as given in [37]:

$$\mathrm{Ga(g)} + \frac{1}{2}\mathrm{As_2(g)} \rightleftharpoons \mathrm{GaAs(s)} \quad \text{and} \tag{2.3}$$

$$2\mathrm{As_2(g)} \rightleftharpoons \mathrm{As_4(g)}, \tag{2.4}$$

where Ga(g) is a Ga molecule in the gas phase, $\mathrm{As_2(g)}$ and $\mathrm{As_4(g)}$ are arsenic dimeric and tetrameric molecules in the gas phase respectively, and GaAs(s) is a GaAs molecule in the solid phase. We assume that the GaAs concentration in the solid state is very close to unity (the maximum deviation approaches 10^{-4} at high temperatures).

The equilibrium constants are defined as:

$$K_i = \exp\left(\frac{\Delta S}{k}\right) \exp\left(\frac{-\Delta H}{kT}\right) = K^{0i} \exp\left(\frac{-\Delta H}{kT}\right), \tag{2.5}$$

where ΔS is the change in entropy associated with the reaction, and ΔH is the enthalpy of the reaction.

The equation of mass action for the reaction (2.3) is

$$\frac{p_{\mathrm{Ga}} p_{\mathrm{As_2}}^{1/2}}{\alpha_{\mathrm{GaAs}}} = (K_{\mathrm{GaAs}})^{-1} = K^*_{\mathrm{GaAs}}$$

$$= 2.73 \times 10^{11} \exp\left(\frac{-4.72\,\mathrm{eV}}{kT}\right), \tag{2.6}$$

where p_{Ga} and $p_{\mathrm{As_2}}$ are the partial equilibrium pressures of gallium and arsenic at the substrate surface (in atmospheres), α_{GaAs} is the activity of the GaAs in the solid phase (equal to unity for a binary compound), K^*_{GaAs} is the inverse equilibrium constant [62], and kT is measured in electron-volts.

The Ga pressure is at its maximum along the Ga–liquid boundary of GaAs and approximately corresponds at moderate temperatures to Ga pressure over pure Ga. When we shift to the As-rich boundary of the solidus curve, the arsenic pressure increases and, according to (2.6), the equilibrium Ga pressure decreases. The total pressure is given by:

$$p^{\mathrm{T}} = p_{\mathrm{Ga}} + p_{\mathrm{As_2}} = \frac{K^*_{\mathrm{GaAs}}}{p_{\mathrm{As_2}}^{1/2}} + p_{\mathrm{As_2}}. \tag{2.7}$$

According to Fig. 2.2, for the Ga–As system at temperature T_1 there exists a minimum in the total pressure for some particular stoichiometry of the solid phase. The vapor pressure corresponding to this minimum can be derived from a condition given by

$$\frac{\mathrm{d}p^{\mathrm{T}}}{\mathrm{d}p_{\mathrm{As_2}}} = \frac{\mathrm{d}p^{\mathrm{T}}}{\mathrm{d}p_{\mathrm{Ga}}} = 0,$$

2.3 Solid–Vapor Equilibrium for Binary Compounds

which results in

$$p_{As_2} = \frac{p_{Ga}}{2} = \left(\frac{K^*_{GaAs_2}}{4}\right)^{1/3}. \tag{2.8}$$

In the case of free sublimation (i.e. sublimation in a vacuum), this condition corresponds to the case of congruent decomposition of the material. In other words, in case of heating the GaAs substrate in a vacuum in this regime, no Ga droplets will appear on the substrate surface.

When the substrate temperature is increased further, both Ga and As pressure at the Ga-liquid boundary increase. At the same time, the As pressure increases faster with temperature, and at some temperature $T_2 > T_1$ the arsenic pressure becomes larger than the Ga pressure in the whole field within the solidus curve. In practice, this means that if the sublimation temperature is below T_2, the state of the system upon evaporation is described by the condition of the minimum total pressure, which corresponds to the case of congruent sublimation ($p^s_{As_2} = \frac{1}{2} p^s_{Ga}$). If the substrate temperature exceeds T_2, the minimum in the total pressure disappears, and the material decomposes in a noncongruent way, when precipitates of the other phase (Ga) are formed on the surface. The critical temperature that separates these two temperature regimes can be derived from (2.1) and (2.6) and is called "the temperature of maximum sublimation". For GaAs this temperature is equal to 630 °C, and it is dangerous to anneal the substrate at higher temperatures in view of the surface morphology degradation. The temperatures of maximum sublimation are listed in Table 2.1.

If we apply an arsenic flux and create an externally defined arsenic pressure over the GaAs surface $p^{ext}_{As_2}$ in such a way that $p^{ext}_{As_2} \gg p^s_{As_2}$, then

$$p^{s1}_{Ga} = \frac{K^*_{GaAs}}{\left(p^{ext}_{As_2}\right)^{1/2}} \ll p^s_{Ga}.$$

This means that the flux of Ga atoms going from the substrate will be much weaker in this case [15]. This effect is well known as an effect of the suppression of sublimation [65]. If we now also apply an external flux of Ga atoms in such a way that it compensates for the sublimation effect, the evaporation rate will be equal to zero.

Now we consider the case of growth, when the external Ga flux is larger than the flux of evaporating Ga atoms. Let us assume that the flux of group-III atoms reaching the surface of the substrate corresponds to a pressure p^0_{III} and that p_{III} is the equilibrium partial pressure of group-III vapour at the surface, so that the growth rate v_g is given by:

$$v_g = \frac{gW(p^0_{III} - p_{III})}{\sqrt{2\pi m k T}}, \tag{2.9}$$

where g is the sticking coefficient ($g = 1$), m is the mass of a molecule, k is the Bolzmann constant, W is the volume of a molecule in a growing crystal, and

Table 2.1. The inverse equilibrium constants, according to [104], ($K_{\text{III-V}}$ (S – K)) and according to [62], ($K_{\text{III-V}}$ (K – L)). At low temperatures, III–V materials decompose congruently, i.e. each arsenic (e.g.) atom leaving the surface is accompanied by a gallium atom. T_{subl} is the temperature of maximal sublimation. T^*_{subl} is the temperature of noncongruent dissociation of III–V compounds for $p^0_V = 2 \times 10^{-5}$ Torr. T^* is the temperature, at which the evaporation rate equals 1 ML/s for $p^0_V = 2 \times 10^{-5}$ Torr

Compound	$K_{\text{III-V}}$ (S–K) $K_{\text{III-V}} = p_{\text{III}} p_{V_2}^{1/2}$	$K_{\text{III-V}}$ (K–L) $K_{\text{III-V}} = p_{\text{III}} p_{V_2}^{1/2}$	T_{subl} (°C)	T^*_{subl} (°C)	T^* (°C)
AlP	$1.39 \times 10^{23} \exp(-5.85/kT)$				
AlAs	$2.90 \times 10^{12} \exp(-5.73/kT)$	$1.63 \times 10^{10} \exp(-5.39/kT)$	902	974	900
GaAs	$1.61 \times 10^{10} \exp(-4.60/kT)$	$2.73 \times 10^{11} \exp(-4.72/kT)$	630	723	700
GaP	$2.85 \times 10^{10} \exp(-4.60/kT)$	$2.26 \times 10^{11} \exp(-4.71/kT)$	571	774	704
InAs	$1.09 \times 10^{11} \exp(-4.22/kT)$	$7.76 \times 10^{11} \exp(-4.34/kT)$	508	688	607
InP	$3.76 \times 10^{10} \exp(-3.88/kT)$	$8.34 \times 10^{11} \exp(-4.02/kT)$	268	684	613
InSb[a]	$3.76 \times 10^{10} \exp(-3.88/kT)$				
GaSb[a]	$1.2 \times 10^{11} \exp(-4.61/kT)$				

[a] for these compounds the equilibrium species under MBE growth conditions are tetrameric molecules.

T is the absolute temperature. Ignoring the weak temperature dependence described by the term $T^{-1/2}$, we obtain from (2.9)

$$v_g = t(p^0_{\text{III}} - p_{\text{III}}), \qquad (2.10)$$

where t is a constant.

For the reaction of (2.4), according to [43, 62] we obtain:

$$p_{\text{As}_4} p_{\text{As}_2}^{-2} = K_{\text{As}} = 27.3 \times 10^3 T - 19.8. \qquad (2.11)$$

The given value of K_{As} is also close to that reported in [141]. In the MBE case, the external arsenic flux fixes the total flux of all arsenic atoms arriving to the surface. Neglecting the part of As atoms participating in growth, we obtain $p^T_{\text{As}} = p_{\text{As}_2} + p_{\text{As}_4}$. The partial pressures of the components can be derived in this case according to:

$$p_{\text{As}_2} = \frac{-1 + \sqrt{1 + 4K_{\text{As}} pT}}{2K_{\text{As}}} \qquad (2.12)$$

$$p_{\text{As}_4} = \frac{\left(-1 + \sqrt{1 + 4K_{\text{As}} pT}\right)^2}{4K_{\text{As}}}. \qquad (2.13)$$

In Table 2.2 we present the equilibrium partial pressures of dimeric and tetrameric species at different temperateres and two different total arsenic pressures. We can conclude that, according to thermodynamic predictions,

Table 2.2. Fraction of dimeric molecules in a total flux from the heated substrate surface [67]

Temperature (°C)	Fraction of As dimeric molecules in the total flux from the surface	
	$p^T = 10^{-5}$ Torr	$p^T = 10^{-6}$ Torr
300	0.009	0.025
350	0.0474	0.155
400	0.237	0.566
450	0.652	0.931
500	0.938	0.992
550	0.992	0.999

the arsenic tetrameric molecules must dominate in the flux from the substrate surface at temperatures of 300°C and below, irrespective of the type of arsenic molecules impinging on the surface. At temperatures of 450°C and higher, however, dimeric molecules start to dominate. Exactly this behavior has been discovered experimentally [28], but was at that time interpreted in the frame of kinetic models of dissociative and associative chemisorption.

Along the Ga boundary of the solidus curve shown in Fig. 2.2a, dimeric molecules dominate at all substrate temperatures. The same conclusion follows from Fig. 2.3, which shows the composition of the gaseous phase along the liquidus curve of a Ga–As system. It is important to note that, in case of VPE – and even in case of sufficiently higher substrate temperatures – tetrameric molecules can contribute significantly to the total As pressure ($0.04\,p^T$ at 800°C, $0.27\,p^T$ at 700°C, $p^T = 10^{-1}$ Torr).

According to thermodynamic considerations, the efficiency of the arsenic species must be similar if the total flux of arsenic atoms is the same. Some authors, however, have claimed [119] that the (Al,Ga)As material quality strongly improves when dimeric molecules are used for growth (e.g. where a cracker cell is introduced to decompose As_4 molecules). Alternatively, extremely high-quality 12 nm-thick GaAs–(Al,Ga)As quantum wells with a full-width at half-maximum (FWHM) of the luminescence emission of 0.3 meV (at 2 K) were realized using beams of tetrameric arsenic molecules [57]. Moreover, FWHM in only the 0.15–0.30 meV range is reported for ultrathin InAs insertions in a GaAs matrix deposited at substrate temperatures of 450–480°C using tetrameric As molecules [113,131]. Alternatively again, at very low temperatures it is not possible completely to exclude kinetic limitations for the decomposition of tetrameric molecules. In these growth conditions, the concentration of nonequilibrium point defects becomes very high and the material quality is strongly affected by the growth rate. Significant improve-

14   2. Basics of MBE Growth

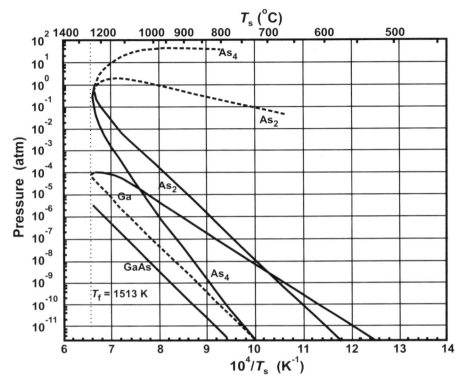

Fig. 2.3. Equilibrium pressures of the components along the liquidus curve for a Ga–As system

ment in quality can usually be obtained by using strongly reduced growth rates [82], which move the system back closer to equilibrium conditions.

In the general case of a nonzero growth rate, the law of conservation of mass gives us:

$$p_{Ga}^0 - p_{Ga} = p_{As}^0 - (4p_{As_4} + 2p_{As_2}),\qquad(2.14)$$

where p_{As}^0 and p_{Ga}^0 correspond to the rate of arrival of the arsenic and gallium atoms respectively on the surface. If $T_s > 400\,°C$ and $p_{As}^0 = 10^{-5}$ Torr, it follows from (2.11) that $p_{As_2} \gg p_{As_4}$, so that if $p_{As}^0 \gg p_{Ga}^0$ then we have $p_{As}^0 \approx 2p_{As_2}$ and

$$v_g = t(p_{Ga}^0 - p_{Ga}) = t\left[p_{Ga}^0 - k_{GaAs}\left(\frac{p_{As}^0}{2}\right)^{-1/2}\right].\qquad(2.15)$$

This dependence describes well the change in the rate of growth of GaAs in molecular beam epitaxy as substrate temperature is varied [39]. The enthalpy of evaporation of GaAs, determined experimentally for MBE conditions by studying the RHEED intensity oscillations due to layer-by-layer evaporation of GaAs, is 4.6 eV [42]. The rate of evaporation at a given temperature is inversely proportional to the square root of the arsenic flux to the substrate, and this relation is in very good quantitative agreement with (2.15). Thus, the thermodynamic description provides exceptionally good agreement with experimental results and no significant kinetic limitations are observed, at least at typical substrate temperatures used for growth.

2.4 Liquid–Solid–Vapor Equilibrium for Binary Compounds

When $p_{As}^0/p_{Ga}^0 \approx 1$ (under the so-called Ga-rich conditions), in other words when growth occurs in the vicinity of the appearance of a second phase (liquid gallium), the value of p_{As_2} from (2.14) is governed by the arsenic vapor pressure along the Ga–GaAs liquidus curve of GaAs [43]. Thus:

$$\left(p_{As_2}^{(Ga_L)}\right)^{1/2} = 9.49 \times 10^5 \times \exp\left(\frac{-1.98}{kT}\right). \tag{2.16}$$

Arsenic precipitates do not form on the GaAs surface at typical MBE growth conditions, because at typically used substrate temperatures the equilibrium arsenic pressure above metallic arsenic ranges from a few Torr to tens of atmospheres, i.e. far beyond typical arsenic beam equivalent pressures used in MBE (10^{-6}–10^{-4} Torr). It follows that the rate of growth given by the thermodynamic model is governed mainly by the rate of arrival of Ga atoms. The excess atoms, which are not used to bind Ga atoms, are re-evaporated. However, the excess arsenic flux determines a point on the phase diagram within the homogeneity region and, consequently, the type and concentration of point defects.

The equilibrium constants according to [62, 104] are given in Table 2.1. The data for [62] is summarized from [93, 121, 122]. There are no major disagreements between the results given by [62, 104] within a factor of 2 or so. The scatter of the results is due to experimental difficulties in evaluation of thermodynamic properties of the components. The interaction parameters in the solid phase according to [118] are given in a Table 2.2.

As has already been noted, at moderate substrate temperatures III–V materials decompose congruently, i.e. each arsenic (e.g.) atom leaving the surface is accompanied by a gallium atom. However, as the substrate temperature increases, the arsenic flux increases and, finally, at some temperature (called the temperature of maximum sublimation T_{subl} [65]) reaches the value corresponding to $p_{As_2}^{1/2(Ga_L)}$. A further increase in substrate temperature will result in noncongruent decomposition of GaAs, and Ga droplets will appear on the surface (T_{subl} for some of the III–V compounds are presented in the Table 2.1 as well). If the substrate is exposed to an external flux of group-V elements

16   2. Basics of MBE Growth

($p_V^0 = 2 \times 10^{-5}$ Torr), the temperature of noncongruent dissociation of III–V compounds can be significantly increased (T_{subl}^* in Table 2.1). The material evaporates congruently, and even the total evaporation rate can be rather high. The temperature at which the evaporation rate equals 1 ML/s for this arsenic pressure is given in Table 2.1 as T^*.

2.5 Particular III–V Materials

2.5.1 AlAs

The reactions between the main components for AlAs growth are given by:

$$\text{Al(g)} + \frac{1}{2}\text{As}_2(\text{g}) \rightleftharpoons \text{AlAs(s)}, \tag{2.17}$$

with the equilibrium constant (see also Table 2.1) given by

$$\frac{p_{\text{Al}} p_{\text{As}_2}^{1/2}}{\alpha_{\text{InAs}}} = (K_{\text{AlAs}})^{-1} = K_{\text{AlAs}}^* = 1.63 \times 10^{10} \exp\left(\frac{-5.39\,\text{eV}}{kT}\right). \tag{2.18}$$

Al pressure along the liquidus curve of the Al–AlAs(s) system is close to Al pressure over the Al-melt at temperatures below 900 °C. Thus we have

$$p_{\text{Al}} = 1.1 \times 10^6 \exp\left(\frac{-3.39\,\text{eV}}{kT}\right). \tag{2.19}$$

Arsenic pressure is correspondingly given as

$$p_{\text{As}_2}^{1/2} = 1.48 \times 10^4 \exp\left(\frac{-2.0\,\text{eV}}{kT}\right). \tag{2.20}$$

The temperature of maximum sublimation is about 900 °C. This high value allows the use of thin AlAs (or AlGaAs) layers as caps for the high-temperature annealing of GaAs. The Al pressure at the Al-rich boundary of the solidus curve at 700 °C is only 2.9×10^{-9} Torr. Thus, even using sub-monolayer (SML) AlAs depositions, followed by high-temperature annealing and AlGaAs overgrowth, allows the fabrication of arrays of quantum wire-like structures (AlAs SML coverage on a vicinal GaAs(100) surface) or quantum dot-like structures (SML islands on a singular GaAs(100) surface).

We should note, however, that the AlAs growth at typical MBE substrate temperatures (550–620 °C) occurs in strongly nonequilibrium conditions. This can result in a high concentration of nonequilibrium point defects, which is particularly dangerous for light-emitting devices. To have good quality AlAs layers, it is usually necessary either to increase the substrate temperature during AlGaAs growth to 700 °C or to reduce the growth rate. The possibility of creating growth conditions closer to equilibrium, even at moderate substrate temperatures, by using a monolayer-thick Ga floating layer will be discussed in Sect. 2.6 below.

2.5.2 InAs

In Fig. 2.4 we show the partial pressures of As_2, As_4, In and InAs species along the In–As liquidus curve [122]. The reactions between the main components for InAs growth are given by

$$\text{In(g)} + \frac{1}{2}\text{As}_2(\text{g}) \rightleftharpoons \text{InAs(s)}, \tag{2.21}$$

with the equilibrium constant (see also Table 2.1) given by

$$\frac{p_{\text{In}} p_{\text{As}_2}^{1/2}}{\alpha_{\text{InAs}}} = (K_{\text{InAs}})^{-1} = K^*_{\text{InAs}} = 7.76 \times 10^{11} \exp\left(\frac{-4.34\,\text{eV}}{kT}\right). \tag{2.22}$$

Indium pressure along the liquidus curve of the In–InAs(s) system can be evaluated as

$$p_{\text{In}} = 1.38 \times 10^5 \exp\left(\frac{2.44\,\text{eV}}{kT}\right). \tag{2.23}$$

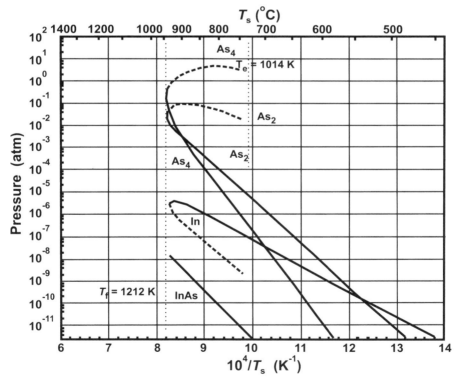

Fig. 2.4. Equilibrium pressures of the components along the liquidus curve for a In–As system

Arsenic pressure is correspondingly given as

$$p_{As_2}^{1/2} = 5.62 \times 10^6 \exp\left(\frac{-1.9\,\text{eV}}{kT}\right). \tag{2.24}$$

The temperature of maximum sublimation is 508 °C, and the arsenic pressure at this temperature is only 1.16×10^{-8} Torr. The temperature of non-congruent decomposition for InAs in the case of external arsenic pressure of 10^{-5} Torr is 688 °C. Indium pressure approaches 10^{-6} Torr (an evaporation rate of about 1 monolayer per second) at the same arsenic pressure at 607 °C. Thus, for typical growth rates of about one monolayer per second, indium evaporation plays a role at temperatures of above 580 °C. Strain effects in the epilayer can further reduce this temperature.

2.5.3 InP

The reactions between the main components for InP growth are given by

$$\text{In(g)} + \frac{1}{2}\text{Ps}_2(\text{g}) \rightleftharpoons \text{InP(s)}, \tag{2.25}$$

with the equilibrium constant (see also Table 2.1) given by

$$\frac{p_{\text{In}} p_{\text{P}_2}^{1/2}}{\alpha_{\text{InP}}} = (K_{\text{InP}})^{-1} = K_{\text{InP}}^* = 8.34 \times 10^{10} \exp\left(\frac{-4.02\,\text{eV}}{kT}\right), \tag{2.26}$$

Indium pressure along the liquidus curve of the In–InP system is:

$$p_{\text{In}} = 1.78 \times 10^5 \exp\left(\frac{-2.45\,\text{eV}}{kT}\right). \tag{2.27}$$

Arsenic pressure is correspondingly given as

$$p_{P_2}^{1/2} = 4.69 \times 10^5 \exp\left(\frac{-1.57\,\text{eV}}{kT}\right). \tag{2.28}$$

Partial components of the total pressure along the In boundary of the phase diagram are shown in Fig. 2.5. The temperature of maximum sublimation for InP is around 270 °C. However, the efficiency of the material decomposition at this temperature is negligible. Indium pressure at this temperature is only 2.7×10^{-15} Torr. At the same time, the temperature of non-congruent decomposition at a phosphorous pressure of 10^{-5} Torr is as high as 684 °C, and the substrate cleaning procedure that uses high-temperature annealing can easily be performed. In contrast, if no phosphorous flux is available (e.g. for (In,Ga,Al)As/InP growth), the substrate cleaning by temperature increase is hardly possible in view of the very low temperature of maximum sublimation. The alternative approaches to produce a clean InP surface (e.g. ion etching, annealing under stabilization by arsenic flux) are to be used when epi-ready substrates are not available.

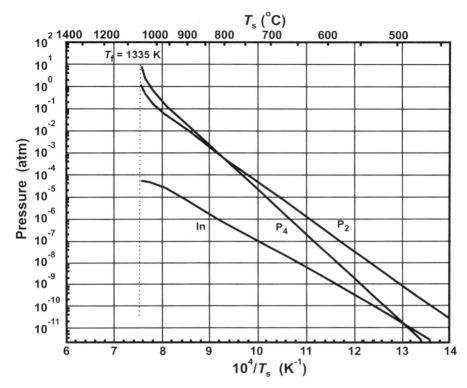

Fig. 2.5. Equilibrium pressures of the components along the metal-rich boundary of the field of solidus of an In–P system

2.5.4 GaP

The equilibrium species along the Ga-rich boundary of the field of solidus for the Ga–P system are shown in Fig. 2.6 [67]. The reactions between the main components for GaP growth are given by

$$\mathrm{Ga(g)} + \frac{1}{2}\mathrm{P_2(g)} \rightleftharpoons \mathrm{GaP(s)}, \tag{2.29}$$

with the equilibrium constant (see also Table 2.1) given by

$$\frac{p_{\mathrm{Ga}} p_{\mathrm{P_2}}^{1/2}}{a_{\mathrm{GaP}}} = (K_{\mathrm{GaP}})^{-1} = K_{\mathrm{GaP}}^* = 2.26 \times 10^{11} \exp\left(\frac{-4.715\,\mathrm{eV}}{kT}\right), \tag{2.30}$$

Ga pressure along the Ga liquidus of the Ga–GaP(s) system is:

$$p_{\mathrm{Ga}} = 3.82 \times 10^5 \exp\left(\frac{-2.75\,\mathrm{eV}}{kT}\right), \tag{2.31}$$

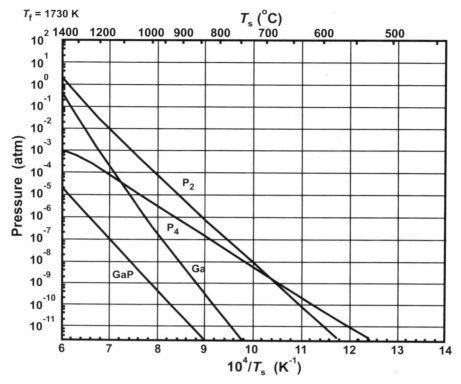

Fig. 2.6. equilibrium pressures of the components along the metal-rich boundary of the field of solidus of a Ga–P system

Phosphorous pressure is correspondingly given as:

$$p_{P_2}^{1/2} = 5.93 \times 10^5 \exp\left(\frac{-1.965\,\text{eV}}{kT}\right). \tag{2.32}$$

The temperature of maximum sublimation is about 671 °C. The gallium pressure at this temperature is 0.73×10^{-6} Torr. This means that GaP substrates can easily be cleaned by high-temperature annealing. The temperature, when Ga pressure approaches 10^{-6} Torr at phosphorous pressure of 10^{-5} Torr, is 704 °C.

2.6 Solid–Vapor Equilibrium for Ternary Compounds

When a solid solution (for example, $\text{Ga}_x\text{In}_{1-x}\text{As}$) is grown by MBE, the activities of GaAs and InAs in the solid phase are smaller than unity:

$$\alpha_{\text{GaAs}} = \gamma_{\text{GaAs}} x = \exp\left[\frac{\Omega_{\text{InAs-GaAs}}(1-x)^2}{kT}\right] x, \tag{2.33}$$

$$\alpha_{\text{InAs}} = \gamma_{\text{InAs}}(1-x) = \exp\left[\frac{\Omega_{\text{InAs-GaAs}}x^2}{kT}\right](1-x), \tag{2.34}$$

where γ_{GaAs} and γ_{InAs} are activity coefficients of GaAs and InAs, respectively, in the solid state. The variable Ω is the interaction parameter [93] – e.g. $\Omega_{\text{AlAs-GaAs}} = 0$, and $\Omega_{\text{InAs-GaAs}} = 0.13\,\text{eV}$. Interaction parameters for some III–V compounds are listed in Table 2.3. Combining (2.31) and (2.34), we obtain

$$\begin{aligned} p_{\text{In}} &= \gamma_{\text{InAs}}(1-x)K_{\text{InAs}}p_{\text{As}_2}^{1/2} \\ &= \exp\left[\frac{\Omega_{\text{InAs-GaAs}}x^2}{kT}\right](1-x)K_{\text{InAs}}p_{\text{As}_2}^{1/2}. \end{aligned} \tag{2.35}$$

The characteristic energy for evaporation (enthalpy value) of $\Delta H = 4.4\,\text{eV}$, found experimentally by [103, 129] from the temperature dependencies of the rates of evaporation of $\text{Ga}_x\text{In}_{1-x}\text{As}$, $\text{Ga}_x\text{In}_{1-x-y}\text{Al}_y\text{As}$ and $\text{Al}_y\text{In}_{1-y}\text{As}$ at typical MBE conditions agrees fairly well with the $4.34\,\text{eV}$ given in Table 2.1. Moreover, the dependence of the evaporation rate on temperature rise and on arsenic pressure [89] agrees qualitatively with the calculations based on (2.35).

For the $\text{Al}_x\text{Ga}_{1-x}\text{As}$ case, $\Omega_{\text{AlAs-GaAs}} = 0$ and the equation for the Ga evaporation rate takes a very simple form:

$$p_{\text{Ga}} = (1-x)K_{\text{GaAs}}p_{\text{As}_2}^{-1/2}, \tag{2.36}$$

Table 2.3. Solid-state interaction parameters in III–V systems

System	$\Omega_{(\text{III-V})-(\text{III-V})}(\text{meV})$
AlAs–GaAs	0
AlAs–InAs	122
AlSb–GaSb	1
AlSb–InSb	63
GaP–GaAs	43
GaP–InP	158
GaAs–GaSb	146
GaAs–InAs	122
GaSb–InSb	80
InP–InAs	25
InAs–InSb	99
GaP–GaN	1254

which shows that at the same values of T_S and p_{As_2} the gallium evaporation rate is higher in the case of GaAs MBE growth as compared with $Al_xGa_{1-x}As$ growth – in agreement with the experimental results [42].

Synthesis of a solid solution containing two group-V elements (for example, $GaAs_xP_{1-x}$) is characterized by mass action equations:

$$p_{Ga}p_{As_2}^{1/2} = \gamma_{GaAs}K_{GaAs}x, \tag{2.37}$$

$$p_{Ga}p_{P_2}^{1/2} = \gamma_{GaP}K_{GaP}(1-x), \tag{2.38}$$

$$K_{As} = p_{As_4}p_{As_2}^{-2}, \tag{2.39}$$

$$K_P = p_{P_4}p_{P_2}^{-2}, \tag{2.40}$$

and the law of conservation of mass:

$$p_{Ga}^0 - p_{Ga} = p_{As}^0 - p_P^0 - (4p_{As_4} + 2p_{As_2} + 4p_{P_4} + 2p_{P_2}). \tag{2.41}$$

Let us assume that $T_S \geq 500\,°C$ and that both p_{As}^0 and $p_P^0 \gg p_{Ga}^0$, which leads to $p_{As_2} \gg p_{As_4}$, $p_{P_2} \gg p_{P_4}$, $p_{As_2} = \frac{1}{2}p_{As}^0$ and $p_{P_2} = \frac{1}{2}p_P^0$, so that (2.37) and (2.38) yield

$$x = \frac{1}{(\gamma_{GaAs}K_{GaAs}/\gamma_{GaAs}K_{GaAs})(p_P^0/p_{As}^0)^{1/2} + 1}. \tag{2.42}$$

The x dependence of activity coefficients does not affect the general trend qualitatively, and for simple estimations we can assume it not to be important. From (2.42) it follows that the composition of the solid phase is independent of the growth rate and of the total flux of group-V molecules, but that it is governed by $(p_P^0/p_{As}^0)^{1/2}$ and the substrate temperature. The results of the calculations based on the thermodynamic model are in good agreement with the experimental data [79].

2.7 Liquid–Solid–Vapor Equilibrium for Ternary Compounds: Surface Segregation of More Volatile Elements

One of essential problems in MBE of III–V compounds and alloys is the surface segregation of the more volatile group-III elements (e.g. Ga in $Al_xGa_{1-x}As$, or In in $Ga_yIn_{1-y}As$ [41, 114]). It is worthy of note that the surface segregation can, to some extent, be considered as a self-organization effect because quasiperiodic corrugations, macroscopic with respect to atomic distance, usually evolve. Thus, proper understanding of this phenomenon brings a key to understanding the main processes related to the spontaneous formation of ordered nanostructures on crystal surfaces.

The surface segregation of a group-III element (Ga in $Al_xGa_{1-x}As$, for instance) was first suggested in [114] as resulting in the appearance of surface corrugation or roughness. This hypothesis was experimentally confirmed [3] by observations using Auger electron microscopy of pronounced Ga accumulation on the surface of $Al_xGa_{1-x}As$ grown by MBE at an intermediate substrate temperature range ($\sim 660\,°C$). Similar effects were observed for indium in the case of $In_xGa_{1-x}As$ and $Al_yIn_{1-y}As$ [77].

From one perspective, surface roughness can make difficult – or even impossible – the creation of abrupt composition profiles, and it may have a deleterious effect on device structure characteristics [114]. From another viewpoint, if it is possible to produce nanoscale quasiperiodic surface corrugations with the aim of constructing surface segregation effects, it is thereby possible to create quantum wires and dots. Laterally quantized structures can be also produced by segregation-driven decomposition of alloy layers, resulting in the formation of compositionally modulated structures.

The influence of growth parameters and conditions on the surface morphology of AlGaAs layers has been extensively discussed by many authors [3, 25, 88, 114, 127]. Most of these authors have pointed out the existence of a so-called "forbidden" substrate temperature range, with its boundaries changing slightly within the regions of 630–640 °C (lower) and 680–700 °C (upper). The $Al_xGa_{1-x}As$ layers grown outside this "forbidden" range were usually found to have a smooth surface. Inside the range, the surface morphology was found to degrade [3, 88, 114] and to be considerably improved by the increase of arsenic pressure, as was shown in [3, 88]. It should also be mentioned that the degree of roughness is strongly affected by the growth rate, the AlAs mole fraction, and the dopants used [3], and also by the (100) substrate misorientation from a singular facet [127].

An attempt to explain the observed results was made in [114]. The authors used a qualitative kinetic model proposed by [34] to describe the behavior of a thin "floating" Ga layer on the AlGaAs surface during growth. However, the physical nature of the processes resulting in segregation of group-III elements or its suppression was not understood. Additionally, the main disadvantage of any kinetic model is the difficulty of determining numerically the model parameters, to relate them to the commonly used growth parameters, to treat the system quantitatively. Using the approach of [51, 52, 62], it is possible to consider the appearance of the liquid phase at typical MBE growth parameters and to apply a completely thermodynamic description.

As described, there exists a minimum possible arsenic equilibrium partial pressure $p_{As_2}^{Ga_L}$, which in the case of GaAs growth is determined by the arsenic pressure over the Ga–GaAs liquidus curve. At $T < 1000\,°C$ this pressure is given by (2.16). According to (2.6), the respective gallium partial pressure is

$$p_{\text{Ga}}^{\text{Ga}_\text{L}} = 2.88 \times 10^5 \exp\left(\frac{-2.74\,\text{eV}}{kT}\right), \tag{2.43}$$

which is essentially the Ga pressure over the liquid gallium. Thus, in the thermodynamic approach the minimum excess arsenic pressure is given by

$$p_{\text{As}_2,\text{exc}} = \frac{1}{2}\left[p_{\text{As}}^0 - (p_{\text{Ga}}^0 - p_{\text{Ga}})\right],$$

which provides the epitaxial growth of GaAs without any liquid phase (Ga droplets) formation on the surface, and which coincides with $p_{\text{As}_2}^{\text{Ga}_\text{L}}$. The value of $p_{\text{As}_2}^{\text{Ga}_\text{L}}$ calculated from (2.16) agrees very well with the experimental arsenic pressure corresponding to the transition from As- to Ga-stabilized conditions under the typical range of substrate temperatures used in MBE [33].

In the case of a ternary compound the arsenic equilibrium pressure over the Al-Ga-Al$_x$Ga$_{1-x}$As liquidus depends (unlike the binary one) on the composition of the solid state x and, respectively, on the composition of the liquid phase that is in equilibrium with it. The general expression for the Ga equilibrium pressure over the complex liquid phase is, according to [93]:

$$p_{\text{Ga}} = p_{\text{Ga}}^* \gamma_{\text{Ga}}[\text{Ga}_\text{L}], \tag{2.44}$$

where $[\text{Ga}_\text{L}]$ and γ_{Ga} are (respectively) the concentration (in molar parts) and the activity coefficient of Ga in the liquid phase. The variable p_{Ga}^* is the Ga equilibrium pressure over the Ga melt, which coincides with $p_{\text{Ga}}^{\text{Ga}_\text{L}}$ [see (2.43)] with a high degree of accuracy.

For the Ga–Al–As liquid phase, which is in equilibrium with the solid solution Al$_x$Ga$_{1-x}$As, the aluminium and arsenic concentrations are negligible in comparison with the gallium concentration at substrate temperatures $T_\text{S} < 800\,^\circ\text{C}$ and solid state compositions $x < 0.8$ [93]:

$$[\text{As}_\text{L}], [\text{Al}_\text{L}] \ll 1, [\text{Ga}_\text{L}] \approx 1. \tag{2.45}$$

The calculation of γ_{Ga} according to [93] for $T_\text{S} = 700\,^\circ\text{C}$ and $x = 0.5$ gives the value of $\gamma_{\text{Ga}} = 1.002$. Thus, taking into account that $\gamma_{\text{Ga}} \approx 1$ in the above temperature and composition range,

$$p_{\text{Ga}}^{\text{L}} = p_{\text{Ga}}^{\text{Ga}_\text{L}} = 2.88 \times 10^5 \exp\left(\frac{-2.74\,\text{eV}}{kT}\right). \tag{2.46}$$

The equilibrium equation for the reaction of GaAs formation in the case of Al$_x$Ga$_{1-x}$As is written as follows:

$$p_{\text{Ga}} p_{\text{As}_2}^{1/2} = \gamma_{\text{GaAs}}(1-x) \times 2.73 \times 10^{11} \exp\left(\frac{-4.72\,\text{eV}}{kT}\right), \tag{2.47}$$

where $\gamma_{\text{GaAs}} = 1$ for Al$_x$Ga$_{1-x}$As, as has already been discussed. It follows from (2.46) and (2.47) for Al$_x$Ga$_{1-x}$As alloys that the expression for arsenic equilibrium pressure over the Al–Ga–Al$_x$Ga$_{1-x}$As liquidus curve is:

2.7 Liquid–Solid–Vapor Equilibrium for Ternary Compounds 25

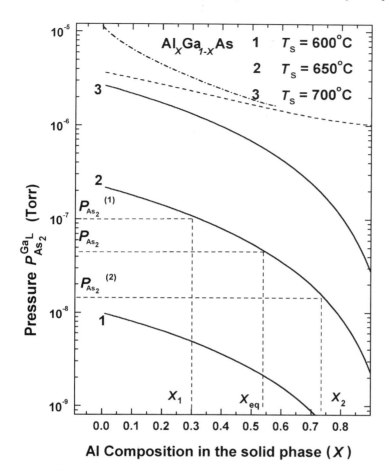

Fig. 2.7. Arsenic equilibrium partial pressure over the Al–GaAs liquid phase, which is in equilibrium with the solid solution $Al_xGa_{1-x}As$, as a function of x at different T_S (*solid lines*); comparison of experimentally measured [33] and theoretically predicted minimum possible excess arsenic pressure of the incident flux that is necessary to obtain good surface morphology at $T_S = 700\,°C$ and a growth rate of $\sim 1\,\mu m/h$ (*dash-dotted line and dashed line, respectively*)

$$\left(p_{As_2}^L\right)^{1/2} = (1-x) \times 2.73 \times 10^{11} \exp(-4.72\mathrm{eV}/kT)/p_{Ga}^{Ga_L}$$
$$= (1-x)\left(p_{As_2}^{Ga_L}\right)^{1/2}, \tag{2.48}$$

that is

$$p_{As_2}^L = (1-x)^2 p_{As_2}^{Ga_L}. \tag{2.49}$$

Figure 2.7 shows the dependencies of arsenic equilibrium partial pressure over the Al–GaAs liquid phase, which is in equilibrium with the solid solution $Al_xGa_{1-x}As$ at different T_S values. The excess arsenic partial pressure determined from the law of conservation of mass is:

$$p_{As_2,exc} = \frac{1}{2}\left[p_{As}^0 - (p_{Ga}^0 - p_{Ga}) - p_{Al}^0\right].\tag{2.50}$$

Variable p_{Ga} can be calculated from (2.47) for given $p_{As_2} = p_{As_2,exc}$; p_{Al} is negligible in comparison with commonly used p_{Al}^0, p_{Ga}^0 at $T_S < 800\,°C$. The temperature dependencies of p_{Ga} at some fixed values of solid-phase compositions x and $p_{As_2,exc}$ are presented in Fig. 2.7 by the dashed and dash-dotted lines. The solid line in this figure corresponds to the gallium partial pressure over the Al–Ga–$Al_xGa_{1-x}As$ liquidus (p_{Ga}^L) [see (2.46)], being the maximum possible Ga pressure. The solid horizontal lines show the beam equivalent pressures of incident Ga flux p_{Ga}^0 at different growth rates.

Let the excess $p_{As_2,exc}$ be less then the equilibrium arsenic pressure over the Ga–GaAs liquidus ($p_{As_2}^{Ga_L}$)$(x = 0)$ at given T_S (see Fig. 2.7, curve 2). In that case, there always exists an equilibrium solid-phase composition x_{eq} corresponding to this $p_{As_2,exc}$. On the other hand, the composition of the bulk-like solid phase is determined mostly by the input gallium and aluminium partial pressures, as

$$x_b = \frac{p_{Al}^0}{[p_{Al}^0 + (p_{Ga}^0 - p_{Ga})]}.\tag{2.51}$$

If, for instance, $x_b = x_1 < x_{eq}$, then the quasiliquid phase (with a "floating" Ga layer) should appear on the surface, because $p_{As_2,exc}$ during growth is less than $(p_{As_2}^L)_1$, corresponding to the Ga–Al–As liquidus for the composition x_1. This quasiliquid phase will be depleted on the arsenic and enriched on the aluminium until it comes to equilibrium with the solid solution with $x = x_{eq}$ determined by $p_{As_2,exc}$. This means that, in the presence of the quasiliquid phase, there appears at least one surface monolayer of $Al_xGa_{1-x}As$ solid solution with the composition $x_{surf} = x_{eq}$ in equilibrium with this liquid phase, while the bulk solid-phase composition is equal to x_1.

The atom exchange reactions between the surface and bulk solid solutions can be neglected in this case, because the diffusion coefficients of gallium and aluminium atoms in the solid phase are very small up to temperatures of $\sim 900\,°C$ [102]. In contrast, the diffusion coefficients of these elements in the liquid phase are as high as $10^{-5}\,cm^2s^{-1}$ [12]. Taking into account also the high rates of exchange reactions between the liquid phase and the surface [12], it is reasonable to conclude that the "liquid-phase-surface-solid-phase-monolayer" equilibrium is realized at all substrate temperatures of interest. In this case, the bulk solid solution grows by capturing the arriving nonequilibrium aluminium, gallium and arsenic atoms, their fast incorporation into the surface solid-phase monolayer, and their reburying by the atoms of the new surface solid-phase monolayer coming to the liquid–solid boundary through the exchange reactions between the liquid phase and the surface.

We should note that any variation of $p_{As_2,exc}$ (provided that $p_{As_2,exc} < p_{As_2}^{Ga_L}$ at given T_S) causes only the respective variations of the liquid phase and surface solid-state monolayer compositions to come to equilibrium with $p_{As_2,exc}$. As $p_{As_2,exc}$ is always equal to the minimum possible excess arsenic pressure for the surface solid-solution monolayer [that is, $p_{As_2,exc} = p_{As_2}^{Ga_L}(x_{eq})$], the liquid phase volume on the surface does not increase.

As was shown above, the quasiliquid phase can exist at any value of $p_{As_2,exc} < p_{As_2}^{Ga_L}$. However, when the excess arsenic pressure is higher than that determined by expression (2.49) for a given bulk solid-solution composition x (see Fig. 2.7, the case of $x_b = x_2$), the situation with $x_{surf} = x_2$ (the absence of a Ga segregation layer) may also come to equilibrium, since $p_{As_2,exc}$ is higher than the pressure corresponding to the Al–Ga–As liquidus boundary for a given composition. The system that has a tendency to increase its area with time will be stable during growth. At the same arsenic excess pressure, the gallium equilibrium pressure over the liquid phase is higher than that over the solid $Al_xGa_{1-x}As$ (see Fig. 2.8). It means that Ga vapor over the liquid-phase regions is supersaturated with respect to the free $Al_xGa_{1-x}As$ surface regions and, if the gallium flux from the gallium cell continues, it will evidently give rise to the condensation of atoms on a free alloy surface.

Similar effects are very well known in quasi-equilibrium growth. The system exhibits a tendency to counteract any change that is forced to occur. In the case of a sublimation, the system approaches the state with the lowest possible equilibrium pressure (see e.g. [65]), and once growth occurs it shifts to the state with the highest equilibrium pressure (evaporation rate) to counteract any growth, if such a state exists. Thus it is possible to conclude that a quasiliquid phase of segregated Ga atoms in thermodynamic equilibrium, formed at any $p_{As_2,exc} < p_{As_2}^{Ga_L}$ and solid-solution composition $x_b < 0.8$, is rather thin and and stable. At growth interruptions, on the contrary, the liquid phase will evaporate faster, and free surface conditions will be stable.

Transitions involving two phases coexisting on the surface usually involve the hysteresis of the transition temperature or arsenic pressure with respect to the direction of the transition [86]. Assuming the interaction between the two phases, there should also be a range of growth conditions when two phases coexist on the surface. This fact was confirmed by tuning the measured enthalpy of Ga evaporation between the values near-characteristic of GaAs evaporation and those characteristic of pure gallium evaporation by changing the average composition of $Al_xGa_{1-x}As$ grown by MBE [26]. This observation indicates that the system is principally mixed, and clearly supports the existence of gallium-covered and gallium-free surface phases. Moreover, different surface phases can potentially form laterally ordered structures [75, 76].

It is also important to underline that the quasiliquid phase does not appear instantly. The surface regions with the Ga segregation layer arise as a result of statistical fluctuations of Ga incident flux. After some critical size of island is reached, the island will increase its size via ripening. Since the gal-

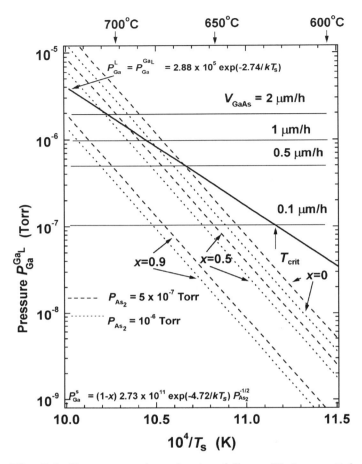

Fig. 2.8. Temperature dependencies of Ga equilibrium partial pressure: over the Al–Ga–Al$_x$Ga$_{1-x}$As liquidus (*solid line*); over the free surface Al$_x$Ga$_{1-x}$As solid phase at different x and $p_{\text{As}_2,\text{exc}}$ (*dashed and dash-dotted lines*). Solid horizontal lines correspond to Ga beam equivalent pressures of the incident Ga flux at different growth rates

lium equilibrium pressure, and thus the Ga flux from the surface, is maximum over these regions (see Fig. 2.8), the GaAs component of the Al$_x$Ga$_{1-x}$As growth rate for these regions is lower and the solid-solution composition x_b is higher, as follows from the law of conservation of mass [see (2.51)] Furthermore, the higher growth rate in the regions free from the segregated gallium atoms will prevent the quasiliquid phase from further propagation over the substrate surface, giving rise to the formation of localized islands of liquid phase.

It is clear that the formation of the quasiliquid phase can cause composition and thickness nonuniformity over the substrate surface and hence be the origin of the $Al_xGa_{1-x}As$ surface morphology degradation.

The growth rate difference between the surface regions with and without a Ga quasiliquid phase is strongly affected by x_b. The increase of x_b decreases the gallium pressure over the free surface regions p_{Ga}^S as compared with p_{Ga}^L (see Fig. 2.8, dash-dotted or dashed lines and solid line) and hence increases the growth rate difference. The higher the latter, the larger the surface roughness for the same growth time. This result agrees well with those obtained experimentally and given in [3], at $T_S = 650\,°C$ and a constant $Al_xGa_{1-x}As$ growth rate, where the increase of roughness degree versus x_b (up to $x_b \sim 0.5$) was measured.

Thus to prevent the formation of a Ga segregation layer during the growth of $Al_xGa_{1-x}As$, it is necessary either (a) to increase the excess arsenic pressure above its equilibrium partial pressure over the Ga–GaAs liquidus at given T_S (As–stabilized growth conditions), that is, $p_{As_2,exc} = \frac{1}{2}\left[p_{As}^0 - (p_{Ga}^0 - p_{Ga}) - p_{Al}^0\right] > p_{As_2}^{Ga_L} = p_{As_2}^L(x=0)$, or (b) to use the Ga beam equivalent pressure p_{Ga}^0 lower than the gallium equilibrium partial pressure along the Ga–GaAs liquidus $p_{Ga}^{Ga-GaAs}$ (but still higher than p_{Ga}^S), when the Ga "floating" layer cannot appear because of the evaporation (see Fig. 2.2b).

The interception points of the horizontal lines and p_{Ga}^L temperature dependence in Fig. 2.8 corresponds to the critical temperatures at which the Ga segregation layer begins to evaporate intensively. These critical T_S values determine the upper temperature boundary of the so-called "forbidden" range at the given growth rate of the GaAs component of $Al_xGa_{1-x}As$. As can be seen from Fig. 2.8, at a typical growth rate of $V_{GaAs} \sim 1\,\mu m/h$ ($p_{Ga}^0 \sim 10^{-6}$ Torr), the critical T_S is equal to $\sim 680\,°C$, which is evidence of rather good agreement between the theory and experimental data [51]. It is clear also that the lower total $Al_xGa_{1-x}As$ growth rate at constant x or the increase of x at constant $Al_xGa_{1-x}As$ growth rate resulting in a lower GaAs growth rate component should reduce the critical temperature. At constant substrate temperature, the smoothing of the surface was clearly observed [3] at a total $Al_xGa_{1-x}As$ growth rate decrease from 2.1 to 0.5 µm/h and at an Al content x_b increase higher than 0.5 at constant total growth rate.

It should be noted that, at high $T_S > T_{S,crit}$, when the quasiliquid phase cannot arise because of evaporation, we have a simple case of two-phase (vapor–solid) interaction, with the dependencies of $Al_xGa_{1-x}As$ growth rate and composition described well by expressions (2.6), (2.47) and (2.51). The composition of the growing solid solution is always $x_b > x_{eq}$ in this growth regime. Further, under these conditions ($T_S > T_{S,crit}$), according to (2.49), the arsenic pressure that is necessary to grow $Al_xGa_{1-x}As$ layers with a smooth morphology can be much lower than that for GaAs growth (see Fig. 2.7).

If the substrate temperature T_S is high enough, but less than $T_{S,\text{crit}}$ at a given growth rate ($p_{\text{Ga}}^L < p_{\text{Ga}}^0$), when the quasiliquid phase is not desorbed then the use of rather low $p_{\text{As}_2,\text{exc}}$ may result in the gallium equilibrium partial pressure over the Ga-free $\text{Al}_x\text{Ga}_{1-x}\text{As}$ surface (p_{Ga}^S) becoming comparable with p_{Ga}^L. In this case, there will be no noticeable difference between the GaAs growth rate components in the surface regions with liquid phase and without it (see Fig. 2.8). Therefore, $\text{Al}_x\text{Ga}_{1-x}\text{As}$ layers grown under these conditions for a reasonable growth time can have a smooth surface morphology. This effect was experimentally observed [3] at $T_S = 680\,°\text{C}$, $\nu_{\text{AlGaAs}} = 2.1\,\mu\text{m/h}$ and $x = 0.3$.

Figure 2.9 shows the calculated temperature dependencies of arsenic equilibrium partial pressure over the Al–Ga–As liquid phase, which is in equilibrium with $\text{Al}_x\text{Ga}_{1-x}\text{As}$ solid solution [dashed lines, see (2.49)]. The dotted line is that obtained experimentally [33] as the minimum possible arsenic beam equivalent pressure corresponding to the transition from As- to Ga-stabilized conditions for undoped GaAs grown by MBE. The temperature dependencies of minimum possible arsenic beam equivalent pressure at which the Al–Ga–As quasiliquid phase does not appear in the frame of the model proposed are presented in Fig. 2.9 by solid lines corresponding to different growth rates. These last dependencies are calculated using the expression:

$$p_{\text{As}_2}^0 = \frac{1}{2}\left[(p_{\text{Ga}}^0 - p_{\text{Ga}}) + p_{\text{Al}}^0\right] + \left(p_{\text{As}_2}^{\text{Ga}_L}\right)^*, \quad (2.52)$$

where $(p_{\text{As}_2}^{\text{Ga}_L})^*$ is taken from the experimental curve (dotted line). As is seen from Fig. 2.9, the effect of an increase in $(p_{\text{As}_2}^{\text{Ga}_L})^*$ becomes noticeable within the substrate temperature range of 630–640 °C, which is in good agreement with the lower boundary of the T_S "forbidden" range. It is clear that the larger the solid-solution growth rate at the same V/III flux ratio, the larger $p_{\text{As}_2,\text{exc}}$, and hence the higher the lower "forbidden" range boundary. Additionally, the hysteresis effects, more pronounced at high growth rates, can increase the critical arsenic flux further. Moreover, according to the current thermodynamic model [50, 52, 62], very low solid-solution growth rates and high V/III ratio should result in the disappearance of the T_S "forbidden" range because of the merging of its upper and lower boundaries (see Figs. 2.8 and 2.9).

The shaded regions in Fig. 2.9 show the growth condition ranges within which, according to the thermodynamic analysis, the Ga segregation layer should appear at growth rates of $2.1\,\mu\text{m/h}$ and $0.7\,\mu\text{m/h}$. The solid circles are the experimental data [3], corresponding to rough and smooth morphology of $\text{Al}_x\text{Ga}_{1-x}\text{As}$ layers respectively. The experimental results of [50] are also shown by solid (rough morphology) and open (smooth morphology) squares. In the latter case, before the growth of each $1\,\mu\text{m}$ $\text{Al}_x\text{Ga}_{1-x}\text{As}$ layer under the growth conditions presented in Fig. 2.9, the $0.5\,\mu\text{m}$ GaAs buffer layer with thin ($\sim 300\,\text{Å}$) AlGaAs layer on top were grown at $T_S = 620\,°\text{C}$. The variation

2.7 Liquid–Solid–Vapor Equilibrium for Ternary Compounds 31

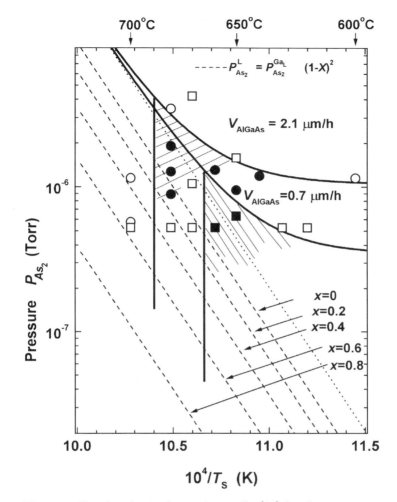

Fig. 2.9. Results obtained experimentally [33] for the minimum possible arsenic beam equivalent pressure corresponding to the transition from As- to Ga-stabilized conditions for undoped GaAs grown by MBE (*dotted line*). The calculated temperature dependences of: arsenic equilibrium partial pressure over the Al–Ga–Al$_x$Ga$_{1-x}$As liquidus at different x (*dashed lines*) and minimum possible arsenic equivalent pressure that is necessary to provide As-stabilized growth conditions at different alloy growth rates of of 2.1 μm/h and 0.7 μm/h (*solid lines*). Comparison of theoretically predicted (*shaded regions*) and experimentally obtained results [3] (*solid and open circles for rough and good morphology, respectively*) are given. The arsenic beam equivalent pressure for the layers grown at 600 °C is believed to be equal to 10^{-6} Torr. Solid and open squares for rough and flat surface morphology, respectively, are experimental results of [50]

of growth parameters (T_S and $p_{As_2,exc}$) for the following 1 μm $Al_xGa_{1-x}As$ layer growth was performed during the growth interruption at this protecting $Al_xGa_{1-x}As$ surface under the arsenic incident flux.

Another way of improving the growth morphology in the "forbidden" temperature range is a predeposition of several Ga monolayers immediately before the growth of any $Al_xGa_{1-x}As$, i.e. when the quasiliquid phase is just beginning to form. Growth occurs through this quasiliquid thermodynamically stable and uniform Ga floating layer ("floating layer epitaxy", as it was called in [52]). Those authors reported a marked improvement in surface morphology, a smaller width of luminescence peak, and a marked increase in electron mobility for the layers grown in such conditions. The growth through the intentionally formed group-III-element quasiliquid phase named "virtual surfactant epitaxy" has also been reported [124, 125].

The case of indium segregation [27, 60, 87] on the surface of $Ga_xIn_{1-x}As$ and $Al_yIn_{1-y}As$ can be considered in a similar way. Results of a suitable thermodynamic analysis predict the possibility of indium surface segregation within the temperature range of 530–600 °C at the growth rate of $\sim 1\,\mu m/h$, and those results are in satisfactory agreement with observed experimental data. For proper treatment, it will be necessary to include the large lattice mismatch for different x compositions that are characteristic of these systems [129]. The effect of strain on phase equilibria and segregation effects will be considered in Chap. 4.

3. Doping and Impurity Segregation Effects in MBE

3.1 Point-Defect Equilibria in MBE

A change in the growth conditions is known to result in a change in a concentration of point defects during growth (see e.g. [65]). There exist no reliable experimental data on point-defect equilibria for MBE GaAs growth. The results obtained in VPE, and particularly in LPE can, however, be used because the growth conditions characteristic of liquid phase epitaxy are known to be close to that used in MBE. In the case of surface segregation effects, the MBE growth conditions correspond to LPE, as was described in the previous chapter. According to [43], the main point defects in LPE and VPE growth are charged arsenic interstitials (As_i^+) and arsenic vacancies (V_{as}^+). The corresponding reactions are:

$$As_{As} + V_i \Leftrightarrow As_i + V_{As} \qquad \Delta H_{fa} \qquad (3.1)$$

$$0 \Leftrightarrow e^- + h^+ \qquad \Delta E_{cv} \qquad (3.2)$$

$$As_i \Leftrightarrow As_i^+ + e^- \qquad \Delta E_{ai} \qquad (3.3)$$

$$V_{As} \Leftrightarrow V_{As}^+ + e^- \qquad \Delta E_{av} \qquad (3.4)$$

$$\tfrac{1}{2} As_2(g) + V_i \Leftrightarrow As_i \qquad \Delta H_{As_2 i} \qquad (3.5)$$

$$0 \Leftrightarrow V_{Ga} + V_{As} \qquad \Delta H_S \qquad (3.6)$$

In the above equations: ΔE_i and ΔH_i are energies and enthalpies of reactions, respectively; e^- and h^+ denote electron and hole, respectively; As_{As} denotes an arsenic atom in the arsenic sublattice; and V_{As} and V_{Ga} are arsenic and gallium vacancies, respectively.

The corresponding mass action equations are thus:

$$[As_i][V_{As}] = K_{fa_i}, \qquad (3.7)$$

$$\gamma_n \gamma_p n p = K_{cv_i}, \qquad (3.8)$$

$$\frac{\gamma_n n [V_{As}^+]}{[A_{As}]} = K_{av_i}, \qquad (3.9)$$

$$\frac{\gamma_n n \left[\mathrm{As}_i^+\right]}{[\mathrm{As}_i]} = K_{\mathrm{ai}_i} , \qquad (3.10)$$

$$[\mathrm{As}_i] \, p_{\mathrm{As}_2}^{1/2} = K_{\mathrm{As}_2 i_i} , \qquad (3.11)$$

$$[\mathrm{V}_{\mathrm{As}}] [\mathrm{V}_{\mathrm{Ga}}] = K_{\mathrm{s}} , \qquad (3.12)$$

where square brackets denote concentrations in molar parts, and

$$n = [\mathrm{e}^-] , \quad p = [\mathrm{h}^+] . \qquad (3.13)$$

To calculate concentrations of the main point defects, it is necesary also to include the electroneutrality condition:

$$\left[\mathrm{V}_{\mathrm{As}}^+\right] + \left[\mathrm{As}_i^+\right] + p = n . \qquad (3.14)$$

According to [43], the equilibrium constants are:

$$K_{\mathrm{cv}} = 1.00 \times 10^{-12} T^3 \exp\left(\frac{-1.62}{kT}\right) , \qquad (3.15)$$

$$K_{\mathrm{fa}} = 2.92 \times 10^6 \exp\left(\frac{-4.845}{kT}\right) , \qquad (3.16)$$

$$K_{\mathrm{av}} = 442 \exp\left(\frac{-0.27}{kT}\right) , \qquad (3.17)$$

$$K_{\mathrm{As}_2 i} = 16.4 \exp\left(\frac{-1.125}{kT}\right) . \qquad (3.18)$$

In the case of doped material, it is necessary to include dopant-related terms in the electroneutrality condition:

$$\left[\mathrm{D}_\Sigma^+\right] + \left[\mathrm{V}_{\mathrm{As}}^+\right] + \left[\mathrm{As}_i^+\right] + p = n + \left[\mathrm{A}_\Sigma^-\right] , \qquad (3.19)$$

where $\left[\mathrm{D}_\Sigma^+\right]$ and $\left[\mathrm{A}_\Sigma^-\right]$ are the sum concentrations of ionized donor and acceptor impurities, respectively. For relatively low arsenic overpressures typically used in MBE, calculations according to (3.1)–(3.19) using the data for the corresponding equilibrium constants show [43, 52, 62, 67] that the main point defects at typical MBE conditions are indeed ionized arsenic vacancies $\left[\mathrm{V}_{\mathrm{As}}^+\right]$. Their concentration can approach $\sim 10^{18}$ cm^{-3} and even higher values at sufficiently high growth temperatures. The results of the calculations, assuming effective arsenic pressure during growth of 10^{-5} Torr, are presented in Fig. 3.1. Lower arsenic pressures result in somewhat higher concentrations of arsenic vacancies.

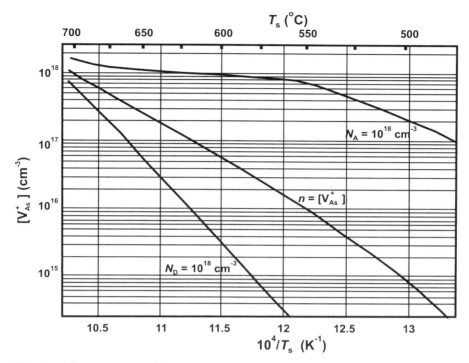

Fig. 3.1. Concentration of the major point defects in MBE of GaAs charged arsenic vacancies $[V_{As}^+]$ for different substrate temperatures and doping concentrations. Effective arsenic pressure during growth, $p_{As} = 10^{-5}$ Torr

It is interesting to note, that, according to Fig. 3.1, in the case of intentional doping with donor impurity, the concentration of ionized vacancies can be rather low, while intentional p-type doping can dramatically increase the V_{As}^+ concentration. The effect of a strong increase in the solubility of charged defects in the presence of high concentration of defects of the opposite sign is, indeed, a well known phenomenon [65]. The concentration of charged vacancies drops quickly with substrate temperature decrease. It can be also assumed that, for reasonable cooling rates, the point defects of different type annihilate one another, or associates of point defects are formed [43]. Thus, the concentration of remaining point defects at room temperature is assumed to be negligible.

We do not consider here the influence of nonequilibrium point defects, which can be formed at very low substrate temperatures due to kinetic limitations.

3.2 Impurity Incorporation in MBE

3.2.1 General Consideration

Let us consider the doping mechanism on an example of the incorporation of an acceptor impurity atom (A) from the gaseous phase A(g) into the gallium sublattice of GaAs. The chemical reactions are given by:

$$A(g) + V_{Ga} \Leftrightarrow A_{Ga}^- + h^+ , \tag{3.20}$$

$$\frac{[A_{Ga}^-] p}{[V_{Ga}] p_A} = K_A , \tag{3.21}$$

where p_A denotes the equilibrium impurity pressure over the surface of the doped material in atmospheres.

It is also necessary to consider the electrical neutrality equation

$$[V_{As}^+] + p = n + [A_{Ga}^-] , \tag{3.22}$$

and the equation of the conservation of mass

$$p_{Ga}^0 [A_{Ga}^-] = p_A^0 - p_A , \tag{3.23}$$

where p_A^0 corresponds to the flux of the impurity atoms arriving at the surface.

If only a small fraction of impurity atoms from the flux to the surface gets incorporated into the growing film, according to (3.23) the dopant concentration is:

$$[A_{Ga}^-] \approx K_A \frac{p_A^0 [V_{Ga}]}{p} , \tag{3.24}$$

or, taking into account (3.12), (3.7) and (3.11):

$$[A_{Ga}^-] \sim \frac{p_A^0 p_{As_2}^{1/2}}{p} , \tag{3.25}$$

i.e. the impurity concentration in the layer is only weakly dependent on growth rate and is defined mostly by the arsenic pressure, the intensity of the impurity flux to the surface, the substrate temperature, and the hole concentration during growth.

In the other limiting case, when most of the impurity atoms arriving at the surface are incorporated into the growing film (i.e. the equilibrium vapor pressure of the impurity gas over the surface of doped material p_A is small), (3.23) results in:

$$[A_{Ga}^-] \approx \frac{p_A^0}{p_{Ga}^0} , \tag{3.26}$$

i.e. the impurity concentration is controlled by the flux ratio between the impurity flux impinging on the surface and the gallium flux, controlling the growth rate.

The same impurity can behave as the impurity of the first type (where a small fraction of atoms from the impurity flux to the surface is incorporated) or the second type (where most of the atoms are incorporated), depending on the substrate temperature. The most important unintentional impurities in MBE are carbon and oxygen atoms from oxygen- and carbon-containing gases of the background atmosphere, and these behave like impurities of the first type. And it is important to note that the concentration of these impurities in the epitaxial layer for the same substrate temperature and excess arsenic pressure depends on the free carrier concentration in the epitaxial film: incorporation of carbon acceptors proceeds more effectively in epilayers intentionally doped with donor atoms. This also explains why carbon atoms incorporate themselves actively in the near-interface region of inverted (n-type AlGaAs/GaAs) interfaces, while this effect is not observed for normal (i.e. GaAs/n-type AlGaAs) interfaces.

At substrate temperatures above approximately 550 °C, examples of impurities of the first type are Mg [137] and Mn [66] in GaAs, and S, Se and Te in GaAs and InP. S-doping of InP is considered in detail in [2] and is shown to be in strict agreement with the considerations given above. The disadvantage of this kind of impurity is the dependence of the dopant concentration on many parameters (e.g. arsenic pressure, substrate temperature); the advantage is the possibility of creating sharp concentration profiles even for ultrahigh doping levels.

Impurities of the second type principally show a tendency to segregate on the surface, as will be shown in the text following. However, the doping control is much easier in this case (e.g. Si, Sn, Be in GaAs and InP). We note that carbon, when incorporated from an intentionally formed flux of carbon atoms on the surface, behaves as an impurity of the second type because its elemental vapor pressure is very low.

3.2.2 Manganese Doping of GaAs

Let us consider the case of manganese incorporations in GaAs epitaxial layers in MBE. Manganese can be used as a p-type dopant with moderate activation energy of 80–100 meV. On the other hand, manganese can be incorporated into epitaxial layers as an unintentional impurity from the heating of the constructive elements of the MBE apparatus that are made of stainless steel.

It has been observed that in the case of unintentional doping at moderate substrate temperatures ($T_S > 500$ °C), the concentration of Mn atoms is much higher in layers strongly doped with donor impurities in comparison with unintentionally doped samples or samples doped with p-type impurities (e.g. beryllium) [66]. It has also been realized that a decrease in arsenic pressure during growth reduces the incorporation of Mn atoms and, moreover, the transition from As-stabilized to Ga-stabilized growth results in a dramatic decrease in the Mn concentration [20, 46, 66, 67].

In cases of intentionally doped GaAs:Mn samples, all these dependencies are found to be much weaker [21]. At low substrate temperatures, all Mn atoms incorporate into the epilayer [66,67]. Kinetic models can hardly explain this observed behavior. In thermodynamic models using an input A/Ga flux ratio as the main parameter [36], it is also difficult to interpret the significant decrease in Mn concentration with transitions from As- rich to Ga-rich growth conditions, as the flux ratio is changed only weakly in these circumstances.

According to (3.21),

$$[\text{Mn}_{\text{Ga}}^-] = \frac{K_{\text{Mn}} p_{\text{Mn}} [V_{\text{Ga}}]}{p} \quad (3.27)$$

in the case of layers intentionally doped with donor impurity. Assuming $np = K_{\text{cv}}$, $n = N_{\text{D}}$ (see Fig. 3.1), and (3.19), we can obtain

$$[\text{Mn}_{\text{Ga}}^-] = K_{\text{Mn}} K_{\text{cv}} K_{\text{S}} K_{\text{As}_2\text{i}} K_{\text{fa}}^{-1} p_{\text{Mn}} p_{\text{As}_2}^{1/2} N_{\text{D}} . \quad (3.28)$$

Assuming the Mn binding energy in the Ga sublattice [35, 66, 67, 96] to be 1.49 eV, the calculation according to (3.28) results in:

$$[\text{Mn}_{\text{Ga}}^-] \sim \exp\left(\frac{2.29}{kT}\right) . \quad (3.29)$$

According to [66, 67], the Mn concentration in heavily doped GaAs:Sn epilayers decreases with substrate temperature during deposition according to

$$[\text{Mn}_{\text{Ga}}^-] \sim \exp\left(\frac{2.4 \pm 0.1}{kT}\right) , \quad (3.30)$$

and in case of Mn incorporation in epilayers doped with acceptor impurity:

$$[\text{Mn}_{\text{Ga}}^-] = \frac{K_{\text{Mn}} p_{\text{Mn}} [V_{\text{Ga}}] (K_{\text{cv}} + K_{\text{av}} [V_{\text{As}}])}{K_{\text{cv}} N_{\text{A}}} . \quad (3.31)$$

In the case of epilayers grown at the same growth conditions ($T_{\text{S}} = 600\,^\circ\text{C}$, $p_{\text{As}_2} = 10^{-5}$ Torr) but differing with the type of impurity used ($N_{\text{D}} = 10^{18}\text{cm}^{-3}$, $N_{\text{A}} = 10^{18}\text{cm}^{-3}$), it is possible to obtain the relative increase of Mn impurity concentration in intentionally donor-doped layers, according to

$$\frac{[\text{Mn}_{\text{Ga}}^-]_{N_{\text{D}}}}{[\text{Mn}_{\text{Ga}}^-]_{N_{\text{A}}}} = \frac{N_{\text{D}} N_{\text{A}}}{(K_{\text{cv}} + K_{\text{av}} [V_{\text{As}}])} = 230 , \quad (3.32)$$

a result that is in a very good agreement with the observed decrease in Mn concentration [66, 67].

The transition from As-rich to Ga-rich growth conditions results in a strong reduction of the excess (effective) arsenic pressure: from 10^{-6}–10^{-5} Torr to values corresponding to the Ga-liquidus of the GaAs system (see Fig. 2.3). Thus the Mn concentration in unintentionally doped or n-type doped epilayers must decrease. For $T_S = 550\,°C$ for n-doped samples, this decrease is of about two orders of magnitude.

In case of epilayers intentionally doped with Mn, the electroneutrality condition can be written as:

$$[\mathrm{Mn_{Ga}^-}] = [\mathrm{V_{As}^+}] + p \tag{3.33}$$

and

$$[\mathrm{Mn_{Ga}^-}] = \left(p_{\mathrm{Mn}} K_{\mathrm{Mn}} K_{\mathrm{av}} K_{\mathrm{S}} K_{\mathrm{cv}}^{-1} + p_{\mathrm{Mn}} K_{\mathrm{Mn}} [\mathrm{V_{Ga}}]\right)^{1/2}. \tag{3.34}$$

According to (3.34), the increase of p_{As} from 2×10^{-6} Torr to 10^{-5} Torr at $T_S = 600\,°C$ increases the Mn concentration by less than 5%. Mn concentration versus T_s dependence becomes weak and the Mn concentration in epilayers at relatively high substrate temperatures ($T_S > 550\,°C$) should show a square-root dependence on Mn flux intensity. At $T_S < 500\,°C$, the equilibrium component corresponding to the Mn flux from the surface of the doped material can be neglected, and all the Mn atoms from the impinging flux can be considered as incorporated into the epilayer [see (3.26)].

3.2.3 GaAs Doping with Zn, Cd, Pb, Mg

Let us now consider the case of Zn doping of GaAs at moderate substrate temperatures ($T_S < 550\,°C$). Then

$$[\mathrm{Zn_{Ga}^-}] \approx p \tag{3.35}$$

$$[\mathrm{Zn_{Ga}^-}] \approx \sqrt{p_{\mathrm{Zn}} K_{\mathrm{Zn}} [\mathrm{V_{Ga}}]}. \tag{3.36}$$

According to [96], $\Delta H_{\mathrm{Zn}} \approx -1.0\,\mathrm{eV}$. Then from (3.36) it follows that

$$[\mathrm{Zn_{Ga}^-}] \sim \exp\left(\frac{0.1}{kT}\right) p_{\mathrm{As_2}}^{1/4}. \tag{3.37}$$

A weak temperature dependence of zinc concentration in GaAs on substrate temperature is in very good agreement with VPE data [112]. It is also known that, in VPE growth, in order to dope GaAs with Zn to $\sim 10^{19}\,\mathrm{cm}^{-3}$, it is necessary to use zinc vapour pressure of the order $p_{\mathrm{Zn}} \approx 5\times 10^{-2}$ Torr. As the equilibrium zinc concentration should decrease with a decrease in arsenic pressure, and the temperature dependence of the concentration is weak, according to (3.37) it is possible to get a high concentration of Zn atoms in epilayers by having Zn pressure of the order of 1 Torr. This is certainly unacceptable in MBE. Thus, reasonably high concentrations of Zn in MBE GaAs epilayers can rarely be achieved. Even to obtain low zink concentrations of the order of $10^{16}\,\mathrm{cm}^{-3}$ it is necessary to have $p_{\mathrm{Zn}} \sim 10^{-6}$ Torr.

A similar situation can be foreseen for Cd, Pb and Mg impurities [35,67].

3.2.4 GaAs Doping with S, Se, Te

It should be noted that, for impurities of the first type, (see Sect. 3.2.1) there can be a situation when the impurity atoms leave the surface in the form of different molecules, like the ones arriving at the surface (and just as it was also considered for the main arsenic components of As$_2$ and As$_4$ molecules). In these circumstances, the corresponding reaction should be included into the consideration and the equilibrium impurity pressure needs to be defined by the flux of molecules evaporating from the surface.

In a case of sulphur doping of GaAs using S$_2$ or S$_4$ molecular beams [67], for example, it is best to consider together the reactions:

$$\frac{1}{2}S_2 + V_{As} \Leftrightarrow S_{As}^+ + e^-, \qquad (3.38)$$

where S$_2$ molecules are the equilibrium species at reasonably high substrate temperatures. At the same time, it is well known that sulphur can easily react with gallium, forming Ga$_2$S molecules:

$$\frac{1}{2}S_2 + 2GaAs \Leftrightarrow Ga_2S + As_2. \qquad (3.39)$$

Taking into account the equilibrium equation for (3.38) $[S_{As}^+] n / p_{S_2}^{1/2} [V_{As}] = K_{Sd}$ and for (3.39) $p_{Ga_2S} p_{As_2} / p_{S_2}^{1/2} = K_{Ga_2S}$, as well as (3.1)–(3.6), and assuming that the electroneutrality condition $n = [S_{As}^+]$ and the mass conservation is low, we have as a result:

$$p_S^0 - [S_{As}^+](p_{Ga}^0 - p_{Ga} - p_{Ga_2S}) = p_{Ga_2S} + 2p_{S_2},$$

where $[S_{As}^+]$ is the sulphur concentration in the growing film. Assuming $p_{S_2}^0 \approx 2p_{Ga_2S}$, we can then obtain:

$$[S_{As}^+] = \frac{K_{Sd} p_{S_2}^{1/2} [V_{As}]}{n} = K_{Sd}^{1/2} p_{S_2}^{1/4} [V_{As}]^{1/2} \approx p_{As_2}^{1/4} (p_{S_2}^0)^{1/2}. \qquad (3.40)$$

For Ga-rich conditions, it is possible to use a simplified formula for the Ga$_2$S formation [35], as follows:

$$2Ga(l) + \frac{1}{2}S_2(g) \Leftrightarrow Ga_2S(g) \qquad (3.41)$$

where Ga(l) denotes a Ga atom in the liquid phase. This gives us

$$\frac{p_{Ga_2S}}{p_{S_2}^{1/2}[Ga_l]} = K_{Ga_2S}^*, \qquad (3.42)$$

where

$$K_{Ga_2S}^* = 1.78 \times 10^5 \exp\left(\frac{0.95}{kT}\right). \qquad (3.43)$$

If we assume that the total pressure of sulfur-related components is limited by the impinging flux pressure, neglecting the part involved in growth and assuming that in MBE Ga-rich conditions $[\text{Ga}_\text{l}] = 1$, we can write

$$p_S^T = P_{\text{Ga}_2\text{S}} + P_{\text{S}_2}, \tag{3.44}$$

if $p_S^T = 10^{-7}$ Torr, and $T_S = 500\,°\text{C}$, then $P_{\text{S}_2}^{\text{eff}} = 2 \times 10^{-40}$ Torr. This means that the equilibrium S_2 pressure, corresponding to the flux from the surface, is low and the incorporation reaction for the sulphur atoms should be written assuming a Ga_2S equilibrium species.

For VPE conditions, assuming $p_{\text{As}_2} = 10^{-1}$ Torr [43], it is necessary to have a total sulphur input pressure of 10^{-4} Torr to get $[\text{S}_\text{As}^+] = 10^{17} \text{cm}^{-3}$ at $T_S = 750\,°\text{C}$. Assuming arsenic pressure in MBE of 10^{-4} Torr and a total sulphur input pressure of 10^{-8} Torr, the maximum sulphur concentration in the epilayer can be of the order of 5×10^{16} cm^{-3} even at very high substrate temperatures. Assuming that a reduction of arsenic pressure in the VPE conditions to 10^{-4} Torr (close to the Ga–liquidus condition at 750 °C) would result in an increase of sulphur concentration up to 3×10^{18} cm^{-3} [according to (3.40)], an estimate of the temperature dependence can also be made. Assuming [96] $\Delta H_S = 4.8\,\text{eV}$ for sulphur bonding in GaAs, an estimate on the basis of combining (3.40), (3.42) and (3.44) can be made that $[\text{S}_\text{As}^+] \sim \exp(0.9/kT)$.

A reduction of the substrate temperature from 750 °C to 500 °C must result in an increase in concentration by about a factor of 20. Thus, doping levels of about 10^{18} cm^{-3} are quite realistic at low temperatures, even for input sulphur pressures of the order of 10^{-10} Torr. This means that, at sufficiently low substrate temperatures, unity sticking coefficients can be realized for moderate growth rates.

It can be also seen from the above consideration that the impurity flux can result in an etching of the crystal surface, provided that the total impurity flux is high enough and the equilibrium constant for (3.38) is also high enough for the given p_{As_2} and T_S. The effect of this can also be used intentionally for a substrate cleaning procedure, or (for example) for etching in lithography. Comparable processes can be shown also to take place for doping with Se and Te atoms using Se$_4$ and Te$_4$ molecular beams.

3.2.5 GaAs Doping with Amphoteric Impurities: Ge, Si, Sn

For the amphoteric impurities of the second type, thermodynamic analysis provides the opportunity to describe the distribution of atoms between sublattices, as it is possible to show it in the example of GaAs doped with Ge according to the LPE and VPE reactions and the equilibrium constants considered [44]:

$$\text{Ge(g)} + \text{V}_\text{Ga} \Leftrightarrow \text{Ge}_\text{Ga}^+ + \text{e}^-, \tag{3.45}$$

$$\mathrm{Ge(g) + V_{As} \Leftrightarrow Ge_{As}^- + h^+}, \tag{3.46}$$

$$\mathrm{Ge_{Ga}^+ + V_{Ga} \Leftrightarrow Ge_{Ga}V_{Ga}^- + 2h^+}. \tag{3.47}$$

The corresponding equilibrium equations are

$$\frac{[\mathrm{Ge_{Ga}^+}]n}{p_{\mathrm{Ge}}[\mathrm{V_{Ga}}]} = K_{\mathrm{Ge_D}}, \tag{3.48}$$

$$\frac{[\mathrm{Ge_{As}^-}]p}{p_{\mathrm{Ge}}[\mathrm{V_{As}}]} = K_{\mathrm{Ge_A}}, \tag{3.49}$$

$$\frac{[\mathrm{Ge_{Ga}V_{Ga}^-}]p}{[\mathrm{Ge_{Ga}^+}][\mathrm{V_{As}}]} = K_{\mathrm{Ge_v}}, \tag{3.50}$$

where p_{Ge} is the equilibrium Ge pressure over the doped GaAs:Ge.

Combining (3.45), (3.46) and (3.47) and taking into account that $np = K_{\mathrm{cv}}$, we obtain

$$\frac{[\mathrm{Ge_{Ga}^+}]}{[\mathrm{Ge_{As}^-}]} = \frac{K_{\mathrm{Ge_D}} p [\mathrm{V_{Ga}}]}{K_{\mathrm{Ge_A}} n [\mathrm{V_{As}}]} = \frac{K_{\mathrm{Ge_D}} K_{\mathrm{cv}} [\mathrm{V_{Ga}}]^2}{K_{\mathrm{Ge_A}} K_S n^2} \approx \frac{p_{\mathrm{As_2}}}{n^2}. \tag{3.51}$$

According to [44] we also have

$$\frac{[\mathrm{Ge_{As}^-}]}{[\mathrm{Ge_{Ga}^+}]} = 5.87 \times 10^{10} \exp\left(\frac{-1.95}{kT}\right) p_{\mathrm{As_2}}^{-1} n^2 \tag{3.52}$$

and

$$\frac{[\mathrm{Ge_{Ga}V_{Ga}^-}]}{[\mathrm{Ge_{Ga}^+}]} = 4.23 \times 10^{-9} \exp\left(\frac{3.99}{kT}\right) p_{\mathrm{As_2}}^{1/2} n^2. \tag{3.53}$$

With the electroneutrality condition (neglecting the concentration of ionized arsenic vacancies) we also have

$$[\mathrm{Ge_{Ga}^+}] + p = [\mathrm{Ge_{As}^-}] + [\mathrm{Ge_{Ga}V_{Ga}^-}] + n \tag{3.54}$$

and, from the equation for the conservation of mass (neglecting the evaporating Ge atoms) we have

$$[\mathrm{Ge_{Ga}^+}] + [\mathrm{Ge_{As}^-}] + [\mathrm{Ge_{Ga}V_{Ga}^-}] = [\mathrm{Ge_T}] = \frac{P_{\mathrm{Ge}}^0}{P_{\mathrm{Ga}}^0}. \tag{3.55}$$

Within the framework of the model, and assuming conditions for Ga-rich growth, where the arsenic pressure is defined by (2.16) and there are low Ge doping levels, we can obtain:

$$\frac{[\mathrm{Ge_{Ga}\,V_{Ga}^-}]}{[\mathrm{Ge_{Ga}^+}]} \ll 1 \quad \text{and} \quad \frac{[\mathrm{Ge_{As}^-}]}{[\mathrm{Ge_{Ga}^+}]} = 5.4\,. \tag{3.56}$$

This observation is in good agreement with experimental data [36]. The transition to As-rich conditions is accompanied by a strong increase in effective As pressure and, according to (3.52), the probability of incorporation to donor sites also strongly increases.

According to (3.52) and (3.53), at particular growth conditions there exists a maximum electron concentration that can be achieved. This concentration can be estimated from the condition $[\mathrm{Ge_{As}^-}]/[\mathrm{Ge_{Ga}^+}] \approx 1$ and equals (at 500 °C, and with $p_{As} = 10^{-6}$ Torr and $p_{As} = 10^{-5}$ Torr) a value of 6×10^{18} cm^{-3} and 2×10^{19} cm^{-3} respectively. These values are in a very reasonable agreement with the experimental data [81].

Thus, to summarize, we can conclude that by using (3.47) and the equilibrium constants provided by [43] that are summarized from the LPE and VPE experimental data, it is possible to predict that in MBE growth conditions under the As-stabilized growth, Ge behaves as a donor impurity, with the degree of compensation increasing with substrate temperature, and the total Ge concentration increasing with decreasing arsenic flux. The maximum n-type doping concentration, arising at 500 °C and $P_{As}^0 \sim 10^{-5}$ Torr, approaches a value of the order of 10^{18} cm^{-3}, but it can be increased to about 10^{19} cm^{-3} for highest arsenic pressures available for MBE growth. For Ga-rich conditions and, in addition, in the case when a Ge segregation layer is formed (for Ge-rich conditions, impurity segregation effects are considered in the forthcoming paragraphs), germanium behaves as an acceptor impurity. All these results are in an excellent agreement with the MBE experimental data [67, 81].

It can be shown that similar effects play a role for Si and Sn dopants, even though the degree of compensation is much smaller in these cases.

3.3 Impurity Segregation in MBE

Impurity segregation is an important effect that prevents the formation of abrupt doping profiles. On the other hand, impurity atoms can be used as a surfactant to modify the growth mechanism and to have an additional degree of freedom in controlling the surface morphology.

In the early stages of MBE, impurity segregation was considered as a kinetically-controlled phenomenon, arising because of kinetic limitations on impurity incorporation in a growing film [32]. Kinetic models for the segregation of tin in GaAs are most the developed [34, 136]. Alternatively, it has been proposed that the surface segregation of tin can have an equilibrium nature [37, 39].

To get an understanding of the nature of an impurity-segregation process using the thermodynamic approach described in the previous sections, we need to consider the arsenic equilibrium pressure over the Ga–As–Sn liquid phase, which is in equilibrium with GaAs doped with Sn. Within the substrate temperature range of interest ($T_S < 800\,°C$), arsenic concentration in the liquid phase can be neglected [23, 43]:

$$[As_L] \ll 1, [Ga_L] + [Sn_L] \approx 1.$$

The gallium equilibrium partial pressure over the Ga–Sn–GaAs:Sn liquidus can be written as follows [see (2.44)]:

$$p_{Ga}^{(Ga-Sn)_L} = \gamma_{Ga} p_{Ga}^{Ga_L} [Ga_L], \tag{3.57}$$

where γ_{Ga} is the Ga activity coefficient in the liquid phase. The calculation of γ_{Ga} using data provided [93] for $T_S = 700\,°C$ and a tin concentration in GaAs of $\sim 10^{18}\,cm^{-3}$ ($[Sn_L] \sim 0.8$, $[As_L] \sim 0.015$, $[Ga_L] \sim 0.2$ [23, 43]), gives the value $\gamma_{Ga} = 1.04$. Further, taking into account that $\gamma_{Ga} \approx 1$, we obtain from (2.6), (2.16), (2.43), and (3.57) the following expression for the arsenic equilibrium pressure:

$$p_{As_2}^{(Ga-Sn)_L} = p_{As_2}^{Ga_L}(1 - [Sn_L])^2. \tag{3.58}$$

The Ga–Sn–As ternary system can be qualitatively compared to the Al–Ga–As one, provided that Ga and Sn play similar roles to Al and Ga respectively. The essential difference is that the doping impurity (Sn in our case) has a solubility limit, i.e. a maximum possible dopant concentration that can be introduced into a semiconductor material (GaAs) at a given growth temperature before impurity precipitation occurs.

Figure 3.2 represents the temperature dependencies of arsenic equilibrium partial pressure in three states: over the Ga–GaAs liquidus ($p_{As_2}^{Ga_L}$); over the Ga–As–Sn liquid phase, which is in equilibrium with GaAs doped with Sn up to the concentration $\sim 10^{18}\,cm^{-3}$ ($p_{As_2}^{(Ga-Sn)_L}$) and which has been calculated according to (3.58) using the experimental data for $[Sn_L]$ and $[Ga_L]$ from [43]; and over the GaAs saturated solution in tin ($p_{As_2}^{Sn_L}$) [115]. For ultrahigh Sn concentrations in GaAs close to the solubility limit ($\sim 10^{19}\,cm^{-3}$ at $700\,°C$), the calculations according to (3.58) and using the experimental data for $[Sn_L]$ and $[Ga_L]$ from [43]) give even higher arsenic pressures.

By analogy with the gallium segregation case, it is possible to show that the value of $p_{As_2,exc}$ should exceed that of $p_{As_2}^{(Ga-Sn)_L}$ for a given Sn doping concentration in equilibrium conditions. During the growth, however, and in similar fashion to the Ga segregation on the $Al_xGa_{1-x}As$ surface, the system will exhibit a tendency to reach the state with the highest possible Sn vapor pressure, the situation that corresponds to $p_{As_2}^{Sn_L}$. This means that during growth, even for relatively small concentration of impurities, there exists a tendency to segregate on the surface if the equivalent pressure of the impurity flux coming at the substrate during growth exceeds the impurity equilibrium

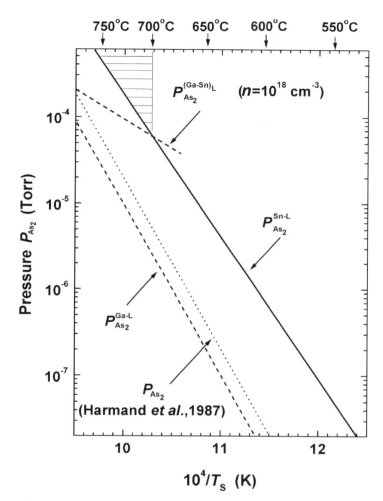

Fig. 3.2. Temperature dependencies of arsenic equilibrium partial pressure: over the Ga–GaAs liquidus (*solid line*); over the Ga–As–Sn liquid phase, which is an equilibrium with GaAs:Sn ($n \sim 10^{18}$ cm^{-3}) (*dashed line*); and over the GaAs saturated solution in Sn (*dash-dotted line*). Dotted line is the same as in Fig. 2.9

pressure over the impurity melt. Alternatively, the Sn concentration at $T_S = 600\,°\mathrm{C}$ and the arsenic pressure corresponding to $p_{As_2}^{Sn_L}$ at this temperature ($\sim 10^{-6}$ Torr) is only 2×10^{17}cm^{-3}. If the Sn doping concentration is higher than this value at $T_S = 600\,°\mathrm{C}$, the tin surface segregation exists at any possible $p_{As_2,\mathrm{exc}}$ value in MBE. It is, however, clear that impurities of the first type (see Sect. 3.2.1) do not segregate on the surface.

It is also important to note that impurity segregation can be observed only if the impurity diffusion coefficient (D) is high enough for the bulk-surface equilibrium to be established during the growth of one monolayer ($D_{\text{crit}} > 10^{-15}\,\text{cm}^{-2}\text{s}^{-1}$ for a growth rate of $\sim 1\,\mu\text{m/h}$) and if it is acceptable that the impurity evaporation from the growth surface can be neglected or be compensated for by an external flux. This is the case for tin in GaAs at moderate doping levels ($\sim 10^{18}\text{cm}^{-3}$) and substrate temperatures ($> 600\,^\circ\text{C}$) [128]. However, in general (for instance, for Be in GaAs), the impurity diffusion is strongly affected by the doping level and the substrate temperature, and D can be lower than the value D_{crit}, leading to kinetic limitations for the formation of an impurity segregation layer. Furthermore, the appearance of an impurity segregation layer must lead to the surface solid-phase layer enrichment with impurity atoms with respect to the bulk (similar to the difference in surface and bulk compositions for the Ga–Al–Al$_x$Ga$_{1-x}$As case).

We can also note that, for Al$_x$Ga$_{1-x}$As:Sn growth by MBE, the thermodynamic description of the surface segregation does not strongly differ from that for the GaAs:Sn except for different Sn solubility limits at the same substrate temperatures, and the expression for arsenic equilibrium pressure over the Al–Ga–As–Sn liquid phase [see (2.49) and (3.58)] can be expressed by

$$p_{\text{As}_2}^{(\text{Al-Ga-Sn})_\text{L}} = \frac{(1-x)^2 p_{\text{As}_2}^{\text{Ga}_\text{L}}}{(1-[\text{Sn}_\text{L}])^2}\,. \tag{3.59}$$

To understand the changes that can be introduced by impurities (or surfactants) in the basic features of epitaxial growth, we shall now consider in more detail the Sn surface segregation in the presence of a Ga segregation layer. The doping of Al$_x$Ga$_{1-x}$As grown under Ga surface-segregation conditions with Sn ([Sn$_\text{S}$] $\sim 10^{19}\,\text{cm}^{-3}$) should result in a variation of the composition of the surface "floating" layer from almost pure Ga to Ga–Sn melt enriched with Sn. Moreover, the thickness of this liquid phase does not increase with time because the surface solid phase monolayer composition x_eq always corresponds to $p_{\text{As}_2,\text{exc}}$, according to (3.59). Furthermore, the Sn surface segregation does not lead to surface morphology degradation. In contrast, the Ga equilibrium pressure over the Al–Ga–Sn quasiliquid phase decreases in accordance with (3.57), resulting in a disappearance of the growth rate difference and, consequently, the composition difference between the surface regions with a segregation layer and without it. Therefore, it becomes possible to grow Al$_x$Ga$_{1-x}$As:Sn layers with smooth surface morphology even within the "forbidden" range of growth conditions (see experimental results of [3]).

Thus the impurity segregation in III–V compounds and alloys can drastically change the situation on the growth surface, giving rise to an increase in $p_{\text{As}_2}^\text{L}$. It is known that there can also exist the segregation of unintentional dopants during III–V MBE growth [83]. On the other hand, it is also known that the surface concentration of unintentional impurities, such as carbon, which can be obtained under nonoptimized growth conditions (for instance at enhanced CO partial pressure) can inhibit the growth of $Al_xGa_{1-x}As$ [98], resulting in a rough surface – carbon can form precipitates, acting as nucleation centres. Therefore much care must be taken when experimental results of different authors are compared because, for the same nominal growth conditions, the real situation on the surface may be very different.

3.4 Interplay Between Impurity Segregation and Diffusion in MBE

We will now consider the interplay between the rate of diffusion in the solid phase and the conditions on the crystal surface. We shall do this by using the case of MBE growth of GaAs:Be as an example. Beryllium is an acceptor impurity in MBE growth of GaAs and $Al_xGa_{1-x}As$. It has long been established [47], [80] that acceptor concentrations as high as $(1-5) \times 10^{19}\,cm^{-3}$ together with perfect surface morphology are routinely obtained under the MBE growth temperatures commonly used (i.e. $580 \pm 20\,°C$). At doping levels up to $(1-4) \times 10^{19}\,cm^{-3}$, the Be diffusion coefficients in both As-grown GaAs:Be layers and annealed ones have been observed to be small enough compared with Sn ones [80] and not to exceed $10^{-14}\,cm^{-2}s^{-1}$ at $700\,°C$ nor $3 \times 10^{-16}\,cm^{-2}s^{-1}$ at $600\,°C$ [84]. That Be diffusion behavior in heavily doped GaAs during post-growth annealing is qualitatively consistent with the simple interstitial–substitutional model that was originally suggested [30] and that was then developed for Zn diffusion in GaAs [132].

However, most of the attempts to increase Be doping levels have resulted in a deterioration of GaAs crystal integrity [24, 94], which appeared to be due to Be surface segregation and precipitation during growth. Furthermore, the onset of fast Be diffusion, with a diffusion coefficient as high as $2 \times 10^{-12}\,cm^{-2}s^{-1}$ at a substrate temperature of $600\,°C$ and doping densities higher than $5 \times 10^{19}\,cm^{-3}$, has been observed under certain growth conditions [94]. Since such unusual behavior of Be diffusion could not be explained in terms of the simple interstitial–substitutional model, [94] proposed a model based on a surface Fermi-level pinning effect in addition. Nevertheless, a net Be acceptor concentration of $(1-2) \times 10^{20}\,cm^{-3}$ with good surface morphology and no noticeable Be redistribution has been demonstrated at $T_S \leq 500\,°C$ [71] and, even at $T_S = 520\,°C$ at high V/III flux ratios [94].

It has also been reported [22] that the increase of Be doping density causes the appearance of a noticeable concentration of nonradiative recombination centers, identified as positively charged Be interstitials. These photoluminescence results have been confirmed by cathodoluminescence (CL) measurements [10], which demonstrated a rapidly decreasing CL intensity for Be doping levels higher than $1.5 \times 10^{19}\,cm^{-3}$.

In addition, [84] reports a long-range assymetric Be redistribution in GaAs towards the growing surface at $600-610\,°C$ at a comparatively low doping level of $5 \times 10^{18}\,\mathrm{cm}^{-3}$, which also cannot be explained by the simple interstitial–substitutional model. The observed Be "carry forward" has tentatively been suggested to be connected with a surface or near-surface accumulation of Be.

The lack of understanding of the primary effects related to Be diffusion and segregation effects has stimulated attempts to develop the thermodynamic models [50, 52]. The important difference in the case of Be surface segregation with respect to Ga and Sn segregation is the difference in the diffusion coefficients in the solid phase. In the case of Ga segregation on a GaAs ($\mathrm{Al}_x\mathrm{Ga}_{1-x}\mathrm{As}$) surface, the diffusion coefficients of Ga and Al atoms in the solid phase are very small, which allows consideration of the equilibrium "liquid-phase surface solid-phase monolayer (SSPM)" at any substrate temperatures of interest without taking into account the SSPM–bulk interaction. Contrary to that, in the case of tin segregation, the diffusion coefficient of Sn atoms in the solid phase is high enough for surface–bulk equilibrium to be established during a one-monolayer growth time during typical growth conditions [136]. However, in general (e.g. for Be in GaAs) this is not the case, and impurity segregation ability depends on whether the rate of atom arrival from the bulk to the surface is sufficiently high. Otherwise, it is necessary to take into consideration the balance of the atomic fluxes (or, in other words, the law of conservation of mass).

The reaction governing Be incorporation into the Ga sublattice of GaAs in MBE is:

$$\mathrm{Be(g)} + \mathrm{V_{Ga}} \Leftrightarrow \mathrm{Be^-_{Ga}} + h^+ \,. \tag{3.60}$$

In the case of three-phase equilibrium on the surface, this reaction can be considered as a superposition of the following reactions:

$$\mathrm{Be(g)} \Leftrightarrow \mathrm{Be(l)}\,, \qquad \Delta H_{\mathrm{surf}}\,, \tag{3.61}$$

$$\mathrm{Be(l)} + \mathrm{V_{Ga}} \Leftrightarrow \mathrm{Be^-_{Ga}} + h^+\,, \qquad \Delta H_{\mathrm{Be}}\,, \tag{3.62}$$

where Be(l) is a Be atom in the quasiliquid phase formed on the surface, and ΔH_{surf} and ΔH_{Be} are the reaction enthalpies.

Applying the law of mass action to (3.62) gives

$$K_{\mathrm{Be}} = \frac{[\mathrm{Be^-_{Ga}}]p}{[\mathrm{Be_L}][\mathrm{V_{Ga}}]} = K_0 \exp\left(-\frac{\Delta H_{\mathrm{Be}}}{kT}\right)\,, \tag{3.63}$$

where ΔH_{Be} can be estimated as the Be bond energy difference in solid and surface quasiliquid phases.

According to [96], it is possible to estimate the single Be–Be, Be–As, Be–Ga bond energies: $E_{\mathrm{Be-Be}} \approx 1.69\,\mathrm{eV}$; $E_{\mathrm{Be-As}} \approx 1.42\,\mathrm{eV}$; $E_{\mathrm{Be-Ga}} \approx 1.32\,\mathrm{eV}$. Taking into account the number of bonds [50, 52], $\Delta H_{\mathrm{Be}} \approx -2.57 \pm 0.27\,\mathrm{eV}$. Then, according to (3.63)

3.4 Interplay Between Impurity Segregation and Diffusion in MBE 49

$$[\text{Be}_{\text{Ga}}^-] = K_0 \exp\left(\frac{2.57 \pm 0.27}{kT}\right) \frac{[\text{Be}_{\text{L}}][\text{V}_{\text{Ga}}]}{p}. \qquad (3.64)$$

Referring also to the electroneutrality equation, we have

$$p + [\text{V}_{\text{As}}^+] = n + [\text{Be}_{\text{Ga}}^-]. \qquad (3.65)$$

The calculation according to [43] for the substrate temperature within the range of 550–650 °C, a Be doping level higher than $10^{18}\,\text{cm}^{-3}$ and arsenic equilibrium pressures of about 10^{-6} Torr shows that n and p are small compared with $[\text{V}_{\text{As}}^+]$ and $[\text{Be}_{\text{Ga}}^-]$, so we can rewrite (3.65) as

$$[\text{V}_{\text{As}}^+] = [\text{Be}_{\text{Ga}}^-]. \qquad (3.66)$$

The ionized arsenic vacancy can be found from (3.4) as

$$[\text{V}_{\text{As}}^+] = K_{\text{av}} \frac{[\text{V}_{\text{As}}]}{n}, \qquad (3.67)$$

and then, taking into account (3.66), (3.12) and $np = K_{\text{cv}}$, we can obtain from (3.64):

$$[\text{Be}_{\text{Ga}}^-]^2 = K_0 K_{\text{av}} K_{\text{S}} K_{\text{cv}}^{-1} \exp\left(\frac{2.57 \pm 0.27}{kT}\right) [\text{Be}_{\text{L}}]. \qquad (3.68)$$

Combining (3.68) and the expressions for K_{av}, K_{S} and K_{cv} from [43] and [133] gives us:

$$[\text{Be}_{\text{Ga}}^-] = K_0^{1/2} \exp\left(\frac{0.28 \pm 0.14}{kT}\right) [\text{Be}_{\text{L}}]^{1/2} \qquad (3.69)$$

It may be concluded from (3.69) that Be solubility in GaAs is not drastically affected by the growth temperature. Since no reliable data on low-temperature LPE growth of GaAs or AlGaAs heavily doped with Be are available, it can only be assumed by analogy with GaAs:Zn and InP:Be that a Be solubility limit in GaAs exists. This is confirmed by the TEM observation of the appearance of Be microprecipitates in heavily doped GaAs MBE layers [22]. Then it follows, from the weak temperature dependence of (3.69) and from the LPE data [31], that the Be solubility limit in GaAs and AlGaAs ($[\text{Be}_{\text{Ga}}^-]_{\text{max}}$) appears to be of the order of $(1-5) \times 10^{20}\,\text{cm}^{-3}$ even at 500 °C.

In this case, by analogy with Sn in GaAs (see the previous section), we can consider the arsenic equilibrium pressure over the Ga–As–Be quasiliquid phase, which is in equilibrium with GaAs heavily doped with Be. Although the melting point of beryllium is 1283 °C, the Be-rich surface segregation layer may be regarded as a quasiliquid phase, which we assume to be thin enough for Be microprecipitates to form. If the Be accumulation exceeds some critical value (approximately one monolayer), Be precipitates are formed on the surface and result in a deterioration of the surface's planar morphology and in the formation of Be precipitates acting as nonradiative recombination centers. For further consideration, we will ignore the arsenic concentration in the liquid phase in view of the exceptionally high arsenic equilibrium pressures over the pure As, i.e.

$$[As_L] \ll 1, [Ga_L] + [Be_L] \approx 1. \tag{3.70}$$

Omitting intermediate calculations here, we can nevertheless derive

$$p_{As_2}^{(Be-Ga)_L} = \frac{p_{As_2}^{Ga_L}}{(1-[Be_L])^2} = \frac{p_{As_2}^{Ga_L}}{\left(1-K'\left[Be_{Ga}^-\right]^2\right)^2}, \tag{3.71}$$

where $p_{As_2}^{Ga_L}$ is the arsenic equilibrium pressure over the Ga–GaAs liquidus of undoped GaAs, as derived from (2.16), the Ga activity coefficient in the liquid phase is taken to be 1 for simplicity, and K' is defined by (3.69). Thus, as it follows from (3.71), at $T_S \sim 600 - 630\,°C$, when $p_{As_2}^{Ga_L}$ becomes noticeable, and at high Be concentration in the quasiliquid phase $[Be_L] \approx 1([Be_{Ga}^-] \approx [Be_{Ga}^-]_{max})$, arsenic equilibrium pressure over the Ga–Be–GaAs:Be liquidus, $p_{As_2}^{Be_L} = (p_{As_2}^{(Be-Ga)_L})_{max}$, becomes much higher than $p_{As_2}^{Ga_L}$. At low temperatures of about $500\,°C$, $p_{As_2}^{(Be-Ga)_L}$ should be small because of the commonly observed As solubility decrease with decreasing temperature in a metal solution.

To consider GaAs:Be surface–bulk equilibrium at different external arsenic pressures and different Be doping levels at zero growth rate, we present in Fig. 3.3 a qualitative dependence of $p_{As_2}^{(Be-Ga)_L}$ on doping densities according to (3.71). Variable $p_{As_2}^{Be_L}$ is the arsenic equilibrium pressure over the GaAs saturated solution in beryllium, which in a simple case corresponds to the Be solubility limit in the solid phase [see (3.69)]. It is clear that, during the growth at $p_{Be}^0/p_{Ga}^0 > \left[Be_{Ga}^-\right]_{max}$ the Be segregation layer formation is inevitable at any excess arsenic pressure, where

$$p_{As_2,exc} = \frac{1}{2}\left[p_{As}^0 - \left(p_{Ga}^0 - p_{Ga}\right)\right]. \tag{3.72}$$

Similar to Sn segregation, it can be shown that the Be segregation layer should not occur if $[Be_{Ga}^-] < [Be_{Ga}^-]_{max}$ and $p_{As_2,exc} > p_{As_2}^{Be_L}$.

It is seen from Fig. 3.3 that there may exist a Ga–Be–GaAs surface equilibrium quasiliquid phase with the arsenic equilibrium pressure over it corresponding to any given beryllium doping level: $[Be_{Ga}^-]_b < [Be_{Ga}^-]_{max}$, and $p_{As_2,exc}^{eq} > p_{As_2}^{(Ga-Be)_L} = f([Be_{Ga}^-]_b)$ – see (3.71). Therefore if, as shown in Fig. 3.3, excess arsenic pressure $p_{As_2,exc} \equiv p_{As_2,exc}^{(1)}$ is less than $p_{As_2}^{eq}$ (but higher than $p_{As_2}^{Ga_L}$), the equilibrium quasiliquid phase with the composition determined by $p_{As_2,exc}^{(1)}$ should occur because $p_{As_2}^{eq}$ is the minimum possible

3.4 Interplay Between Impurity Segregation and Diffusion in MBE 51

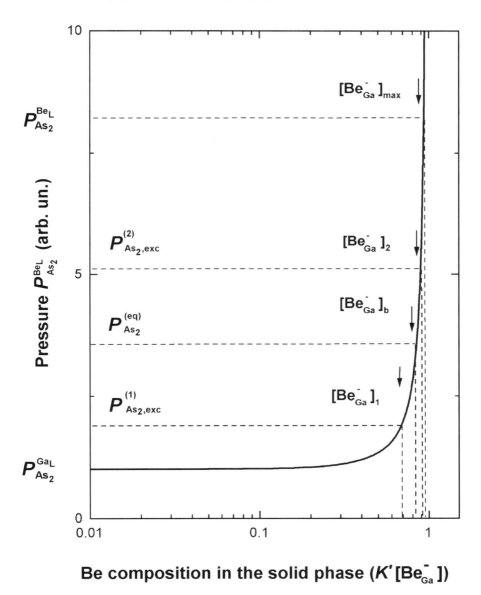

Fig. 3.3. The theoretical dependence of arsenic equilibrium pressure over the Ga–Be–GaAs:Be liquidus on the beryllium concentration in the solid phase, illustrating the mechanism for the formation of a Be segregation layer

arsenic pressure for a given $[\text{Be}_{\text{Ga}}^-]_b$. This effect should result in a depletion of the near-surface GaAs solid-phase layer with beryllium, as compared to the bulk, such that even the Ga–Be segregation layer is formed on the surface $\left([\text{Be}_{\text{Ga}}^-]_1 < [\text{Be}_{\text{Ga}}^-]_b\right)$. In practice, this means that, for given arsenic pressure and a very high Be concentration in the bulk, a segregation layer composed of Ga and Be should inevitably be formed on the surface. Surprisingly, a decrease in the excess arsenic pressure should result in a decrease of Be concentration in the quasiliquid phase and also in a further decrease in the equilibrium Be concentration in the near-surface solid-phase layer, resulting also in a reduced tendency for carry-forward diffusion.

When $p_{\text{As}_2,\text{exc}} = p_{\text{As}_2,\text{exc}}^{(2)} > p_{\text{As}_2}^{\text{eq}}$ (see Fig. 3.3), the near surface GaAs layer, which is in equilibrium with the quasiliquid phase determined by $p_{\text{As}_2,\text{exc}}^{(2)}$, must have a higher Be concentration than that in the bulk ($[\text{Be}_{\text{Ga}}^-]_2 > [\text{Be}_{\text{Ga}}^-]_b$. However, in this case, when $p_{\text{As}_2,\text{exc}}^{(2)} > p_{\text{As}_2}^{\text{eq}}$, the situation with $[\text{Be}_{\text{Ga}}^-]_{\text{surf}} = [\text{Be}_{\text{Ga}}^-]_b$ without a Be segregation layer may also come to equilibrium, since $p_{\text{As}_2,\text{exc}}^{(2)}$ is higher than the pressure corresponding to the Ga–Be–GaAs:Be liquidus boundary for a given bulk concentration. When the free surface regions come into contact with the regions with segregated Be, the latter have a tendency to enlarge the surface (reasonable by analogy with the Ga segregation on the AlGaAs surface). Hence, any accidental fluctuation of the Be surface concentration should result in covering the GaAs:Be surface with the Be segregation layer.

From (3.71) and Fig. 3.3 it is possible to conclude that the dependence of $p_{\text{As}_2}^{(\text{Ga}-\text{Be})_L}$ on $[\text{Be}_{\text{Ga}}^-]$ is very steep only in the vicinity of $[\text{Be}_{\text{Ga}}^-]_{\text{max}}$. So any $p_{\text{As}_2,\text{exc}}$ sufficiently higher than than $p_{\text{As}_2}^{\text{Ga}_L}$ will be in equilibrium with a very high concentration of beryllium in the solid phase. This means that, in practice, for MBE doping of GaAs with Be, when the Be bulk concentration is sufficiently low compared with the Be solubility limit (i.e. $[\text{Be}_{\text{Ga}}^-]_b < (1-5) \times 10^{20} \text{cm}^{-3}$), the Be concentration in the surface layer can be close to $[\text{Be}_{\text{Ga}}^-]_{\text{max}}$.

A beryllium segregation layer can, however, appear on the surface only if the Be diffusion coefficient in the solid state is large enough. As has been proposed and confirmed experimentally by [84,85], the diffusion of Be in GaAs (and AlGaAs) may proceed via highly mobile positively charged Be interstitials (Be_i^+), while the resulting profile of electrically active beryllium atoms, defined by substitutional Be acceptors, is determined by the interaction of the interstitials with gallium vacancies [84] in similar fashion to the case of Zn diffusion in GaAs [30,132]. Thus:

$$\text{Be}_i^+ + V_{\text{Ga}} \Leftrightarrow \text{Be}_{\text{Ga}}^- + 2h^+ . \tag{3.73}$$

The concentration of interstitials is assumed to be small compared with substitutional ones, i.e. $[\text{Be}_i^+] \ll [\text{Be}_{\text{Ga}}^-]$. Applying the law of mass action to (3.73) gives:

3.4 Interplay Between Impurity Segregation and Diffusion in MBE

$$K_{\text{Be}}^{\text{diff}} = \frac{[\text{Be}_{\text{Ga}}^-]p^2}{[\text{Be}_i^+][\text{V}_{\text{Ga}}]}. \tag{3.74}$$

According to [30], the expression for the Be diffusion coefficient should then be written as

$$D_{\text{Be}} \sim \frac{[\text{Be}_{\text{Ga}}^-]}{[\text{Be}_{\text{Ga}}^-]} = \frac{(K_{\text{Be}}^{\text{diff}})^{-1}p^2}{[\text{V}_{\text{Ga}}]}. \tag{3.75}$$

Taking into account (3.66), (3.67) and (3.12) for $[\text{Be}_{\text{Ga}}^-] > 10^{18} \text{cm}^{-3}$, moderate excess arsenic pressures ($\sim 10^{-6}$ Torr) and substrate temperatures in the range of 550–650 °C, we can obtain (all the equilibrium constants being according to [43]):

$$D_{\text{Be}} \approx K_{\text{cv}}^2 K_{\text{As}_2}^i (K_{\text{fa}} K_{\text{av}}^2 K_{\text{S}} K_{\text{Be}}^{\text{diff}})^{-1} p_{\text{As}_2}^{1/2} \left[\text{Be}_{\text{Ga}}^-\right]^2 = K_1 \left[\text{Be}_{\text{Ga}}^-\right]^2, \tag{3.76}$$

where K_1 is the proportionality coefficient depending on temperature and arsenic pressure.

The diffusion coefficient initially increases with an increase in excess arsenic pressure ($\sim p_{\text{As}_2}^{1/2}$). However, any further increase (above $\sim 10^{-5}$ Torr) of $p_{\text{As}_2,\text{exc}}$ results in a decrease of the concentration of ionized arsenic vacancies [see (3.67) and (3.1)–(3.18)] in such a way that, in the electroneutrality equation, the concentration of holes starts to dominate over the concentration of ionized arsenic vacancies:

$$p \approx [\text{Be}_{\text{Ga}}^-]. \tag{3.77}$$

This equation is also true for relatively low arsenic pressures at T_S below 500–550 °C. For this electroneutrality condition, the diffusion coefficient decreases with an increase in arsenic equilibrium pressure ($D_{\text{Be}} \approx p_{\text{As}_2}^{1/2}$). Thus, for very high excess arsenic pressures, a significant decrease of the beryllium diffusion coefficient with $p_{\text{As}_2,\text{exc}}$ can be expected.

Using the Be diffusion coefficients measured experimentally at different doping levels and substrate temperatures, and taking into account (3.76), it is possible to estimate the distance (L) that is reached by the diffusion front in one second after contact of the doped and undoped regions. That distance is given by

$$L = \sqrt{D_{\text{Be}}} \approx [\text{Be}_{\text{Ga}}^-]. \tag{3.78}$$

Let d be the thickness of the GaAs layer grown in one second. If $L > d$, the Be concentration in the surface monolayer reaches $[\text{Be}_{\text{Ga}}^-]_\text{b}$ for its growth time even without any input Be flux. Thus the input Be flux turns out to be excessive and may lead to surface enrichment with beryllium (if this is energetically favorable) and, as a consequence, the formation of a Be segregation layer.

54   3. Doping and Impurity Segregation Effects in MBE

Figure 3.4 presents temperature and total beryllium concentration dependencies [see (3.78)] of the diffusion distance per second (L). These dependencies are plotted using the experimental data of high-resolution secondary-ion mass spectroscopy (SIMS) profile measurements of [85] at $[\text{Be}_{\text{Ga}}^-]_{\text{b}} = 10^{19}\text{cm}^{-3}$ and 10^{18}cm^{-3} (*crosses, solid line as a guide for the eye*) and the data of [94] for MBE at $T_S = 600\,°C$ (*closed circles*), which agrees with the square dependence of the diffusion coefficient on the beryllium concentration (i.e. for concentrations below the onset of ultrafast diffusion). Dashed lines are calculated according (3.78) assuming that the temperature dependence of the diffusion coefficient remains unchanged. We can see a fairly good agreement for the Be diffusion coefficient applying to MBE growth annealing and implantation up to $[\text{Be}_{\text{Ga}}^-]_{\text{b}} = 2.2 \times 10^{19}\text{cm}^{-3}$ with the annealing occurring at $T_S = 900\,°C$ [80] (*open circle*). An estimation of D_{Be} temperature dependence from Fig. 3.4 gives

$$D_{\text{Be}} \approx \exp\left(\frac{-3.0 \pm 0.2}{kT}\right). \qquad (3.79)$$

The horizontal solid lines in Fig. 3.4 show the values of d at different growth rates. Usually the excess arsenic pressure during the growth is higher than $p_{\text{As}_2}^{\text{eq}}$ but less than $p_{\text{As}_2}^{\text{Be}_\text{L}}$ ($p_{\text{As}_2,\text{exc}} = p_{\text{As}_2,\text{exc}}^{(2)}$, see Fig. 3.4). The total Be concentration $[\text{Be}_{\text{Ga}}^-]_{\text{b}}$ is determined by the Be:Ga flux ratio:

$$[\text{Be}_{\text{Ga}}^-]_{\text{b}} = \frac{p_{\text{Be}}^0}{p_{\text{Ga}}^0}. \qquad (3.80)$$

At low $[\text{Be}_{\text{Ga}}^-]_{\text{b}}$ the diffusion coefficients are very low and all the Be atoms arriving onto the growth surface are incorporated as substitutional acceptors on the Ga sites. As $[\text{Be}_{\text{Ga}}^-]_{\text{b}}$ rises, L increases. At last, when $L \geq d$, the Be diffusion from the region beneath the surface towards the growing film becomes significant, the input Be flux becomes excessive, and the quasiliquid phase of segregated Be and Ga atoms appears. It is in equilibrium with the surface solid-phase monolayer (SSPM), with $[\text{Be}_{\text{Ga}}^-] = [\text{Be}_{\text{Ga}}^-]_2 > [\text{Be}_{\text{Ga}}^-]_{\text{b}}$. Since the beryllium concentration at the SSPM is much higher than in the bulk, the Be interstitial concentration $[\text{Be}_i^+]$ in the SSPM drastically increases according to (3.1)–(3.18), (3.73), (3.74) and (3.77). And so we have

$$[\text{Be}_i^+] \approx [\text{Be}_{\text{Ga}}^-]_{\text{b}}^3. \qquad (3.81)$$

Thus, in this case the growth surface acts as an interstitial generation region, stimulating strong Be diffusion into the bulk. This makes it impossible to obtain a high Be concentration in a thin layer, which is confirmed by the data of [94] in which a drastic increase was observed in the Be diffusion coefficient, from $\sim 1.4 \times 10^{-15}\text{cm}^2/\text{s}$ to $2 \times 10^{-12}\text{cm}^2/\text{s}$ with the Be doping level increasing from $3.8 \times 10^{19}\text{cm}^{-3}$ to $5.3 \times 10^{19}\text{cm}^{-3}$ at $600\,°C$ and the

3.4 Interplay Between Impurity Segregation and Diffusion in MBE   55

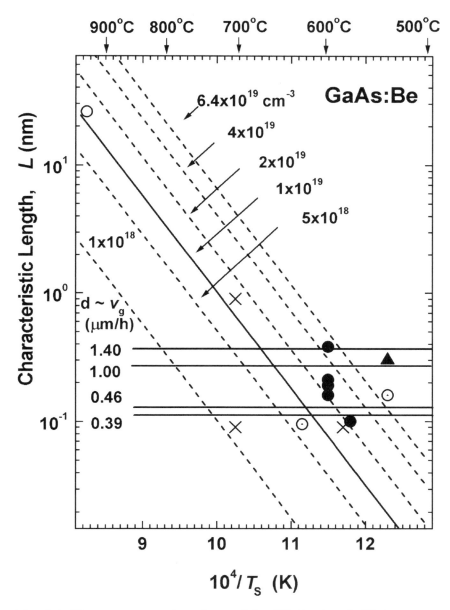

Fig. 3.4. The temperature dependence of the diffusion length (L) in one second at different Be doping levels. (*Solid lines* are the experimental data of [84], [85]; *dashed lines* are calculations for different doping levels; *solid horizontal lines* show the layer thickness formed in one second (d) for different growth rates). According to this figure, the onset of ultrafast Be diffusion, coincident with the formation of a Be segregation layer, occurs when the dotted line for a given temperature and a doping layer is lying higher than the horizontal line for the given growth rate. In other words (for example), ultrafast beryllium diffusion does not occur at $T_S = 550\,°\mathrm{C}$, $[\mathrm{Be}^-_{\mathrm{Ga}}]_\mathrm{b} = 6 \times 10^{19}\,\mathrm{cm}^{-3}$ and a growth rate of $1\,\mathrm{\mu m/h}$, but starts at a growth rate of $0.5\,\mathrm{\mu m/h}$

growth rate of 1.4 μm/h (see Fig. 3.4, *closed circles* and the horizontal line corresponding to $v_G = 1.4\,\mu\text{m/h}$).

In thin heavily Be-doped layers, this effect prevents the Be segregation from forming, because the strong increase of Be in-diffusion flux results in an increase to the thickness of the Be doped layer instead of larger $[\text{Be}_{\text{Ga}}^-]_b$.

For thick heavily doped layers, the Be in-diffusion rate decreases with growth time because of the enrichment of the near-surface layer with Be interstitials:

$$v_D = \frac{L}{t} = \sqrt{\frac{D}{t}}\,. \tag{3.82}$$

Therefore, even at $D_{\text{Be}} = 10^{-12}\,\text{cm}^2/\text{s}$ the Be in-diffusion flux may not provide the depletion of the beryllium-rich surface for a growth rate of ~ 1 monolayer per second after about 20 min of growth. As a result, a Be segregation layer forms. Alternatively, an increase in the concentration of beryllium interstitials ($[\text{Be}_i^+]$) in the growing layer, which are found to behave as nonradiative recombination centers [22], results in a pronounced reduction of the efficiency of luminescence at doping levels corresponding to the condition $L = d$.

Furthermore, as the substrate temperatures used in MBE are small compared with the melting point of beryllium (1223 °C), it is possible for segregated Be atoms to form precipitates of the Be solid phase directly onto the growing surface, resulting in a degradation of the surface morphology. The morphology degradation has indeed been reported [95] to be associated with the onset of fast Be diffusion.

To prevent the formation of the Be segregation layer when $d \leq L$ at a constant growth rate and Be doping level, it is necessary either to reduce significantly the substrate temperature for $p_{\text{As}_2}^{\text{Be}_L}$ to be less then $p_{\text{As}_2,\text{exc}}$, or to increase significantly $p_{\text{As}_2,\text{exc}}$. Very high arsenic pressures can reduce the Be diffusion coefficient and thus also suppress the segregation effect. An improvement in surface morphology by reducing the substrate temperature from 550 °C to 500 °C at the Be concentration of $10^{19}\,\text{cm}^{-3}$ has been reported [71]. In addition, the authors of [95] succeeded in suppressing fast Be diffusion and obtaining a good surface morphology by strongly increasing the As_4/Ga flux ratio. This result was observed at $T_S = 520\,°\text{C}$ for a Be concentration of $10^{19}\,\text{cm}^{-3}$, but not at $T_S = 580\,°\text{C}$. This result agrees well with the theoretical predictions represented in Fig. 3.4.

It is also clear from the model that the Be surface segregation effects could be observed at relatively lower doping levels for the same substrate temperature, if the growth rates are small enough. It has been reported [45] that a transition has been observed from smooth to rough morphology at $2.2 \times 10^{19}\,\text{cm}^{-3}$ at the growth rate of $0.46\,\mu\text{m/h}$ (*closed square* in Fig. 3.4 at $T_S = 570\,°\text{C}$).

Actually, all impurities of the second type (see Sect. 5.2.1) with a concentration-dependent diffusion coefficient (e.g. Sn, Si [128]) should exhibit

3.4 Interplay Between Impurity Segregation and Diffusion in MBE

behavior similar to Be. Although the dependence of the diffusion coefficient on impurity concentration is weaker for them as compared to the Be case, the effect of the onset of fast impurity diffusion has not yet been studied for these impurities.

4. Influence of Strain in the Epitaxial Film on Surface-Phase Equilibria

4.1 MBE Growth of Lattice-Mismatched Binary Compound

Lattice mismatch between the substrate and the epitaxially deposited material strongly affects the whole of the main growth process and can also result in the self-organization of ordered nanostructures on crystal surfaces. We will consider here the influence of strain on solid–vapor and on solid–vapor–liquid equilibrium states in typical MBE conditions for InSb growth on lattice-mismatched substrates.

Such InSb/GaAs and InAs/Si systems represents a case of extreme lattice mismatch of $\Delta a/a = 14.6\%$ and 19.0%, respectively. The modification of the growth-related surface processes is the strongest for these systems, making them of particular interest for fundamental studies of lattice-mismatched growth [54]. From the point of view of practical application, some variability in low electron mass and high room temperature makes InSb attractive for applications in microelectronics, while the smallest bandgap energy among the III–V binary compounds (0.175 eV) makes it attractive as a basic material for producing infrared photodetectors and emitters operating in the window of minimum atmospheric absorption, namely 8–12 µm (see for example [116]).

The properties of InSb films grown on these substrates have been studied in a number of works [7, 14, 17, 18, 70, 91, 92, 139]. Scanning electron microscopy (SEM), transmission electron microscopy (TEM), and differential X-ray diffractometry were usually applied. The electrical and optical characteristics [18] of homo-epitaxial InSb/InSb [1, 91, 134] and hetero-epitaxial InSb/GaAs [17, 91, 92, 135, 139] and InSb/Si [18, 100, 140] have already been studied.

According to [17] and [92], InSb layers with mirror-like surface morphology can be obtained at $T_S < 400\,°\mathrm{C}$ only within a narrow Sb_4/In flux ratio range of around ~ 2.5–$3.5\,\mu m$. At the same time, at $T_S = 420\,°\mathrm{C}$ [134] a perfect surface morphology was obtained even at a Sb_4/In ratio eight times larger than the minimal level applying at low T_S. Significant deterioration of surface morphology, accompanied by the formation of macroscopic indium-rich droplet-like surface structures was observed [17, 18] when InSb/GaAs hetero-epitaxial growth was initiated at $T_S \geq 400\,°\mathrm{C}$. A similar phenomenon was

60   4. Influence of Strain in the Epitaxial Film on Surface-Phase Equilibria

also observed for GaSb/GaAs MBE growth [53] and for InSb/GaAs growth by metal–organic chemical-vapour deposition (MOCVD) [8]. Reference [134] reported the existence of a lower boundary of the T_S range where single crystalline InSb films may be grown by MBE at constant v_G and $\mathrm{Sb}_4/\mathrm{In}$ ratio.

To understand the InSb growth process on highly lattice-mismatched substrates, kinetic studies were carried out [140] on the interaction of Sb_4 and In fluxes with the surface of a heated Si substrate using mass-spectroscopic registration of the fluxes desorbed from the surface at different temperatures. A quantitative thermodynamic analysis has been presented [126] of the InSb phase diagram under MBE growth conditions using an assumption that dimeric Sb molecules are the equilibrium species in the given range of substrate temperatures, total Sb fluxes, similar to the GaAs case, for example. As it was later shown [54], this assumption is not valid and antimony tetramers are the main equilibrium antimony species.

High-quality InSb films having a lowest FWHM value of the X-ray rocking curve and high electron mobility were obtained only in a narrow range of $\mathrm{Sb}_4/\mathrm{In}$ flux ratios, close to In-stabilized growth conditions. Mobility as high as $250\,000\,\mathrm{cm}^2\mathrm{V}^{-1}\mathrm{s}^{-1}$ and $50\,000\,\mathrm{cm}^2\mathrm{V}^{-1}\mathrm{s}^{-1}$ at 77 K were reported for InSb layers on InSb [91] and GaAs substrates [92] respectively. In all other growth conditions, the reported values of the electron mobility were significantly lower.

For the thermodynamic consideration, it is necessary to consider all the equilibrium species for typical growth temperatures and total antimony fluxes. Entropies and heat-of-formation values used in the presented calculations are listed in Table 4.1 according to [54].

Table 4.1. Thermodynamic parameters for the In-Sb system

Component	ΔH^0_{298}(Kcal/mol)	S^0_{298}(Kcal/mol grad)
$\mathrm{Sb}_4(\mathrm{g})$	46.2	80.0
$\mathrm{Sb}_2(\mathrm{g})$	59.8	60.5
$\mathrm{In}(\mathrm{g})$	57.5	41.5
$\mathrm{InSb}(\mathrm{s})$	-7.3	20.82
$\mathrm{Sb}(\mathrm{s})$	0	11.4

First we consider the reaction between antimony dimers and tetramers:

$$2\mathrm{Sb}_2(\mathrm{g}) \Leftrightarrow \mathrm{Sb}_4(\mathrm{g})\,. \tag{4.1}$$

Using the entropy and heat-of-formation values given in Table 4.1, we can show that, for T_S in the range below 530 °C and total antimony pressures $p_{\mathrm{Sb,tot}} = 10^{-7}\text{--}5\times 10^{-6}$ Torr, the equilibrium antimony species are the tetrameric molecules ($> 90\%$ of $p_{\text{Sb,tot}}$). Thus, the Sb$_4$ molecules are expected to dominate the flux from the surface at typical InSb MBE growth conditions ($T_{\text{S}} < 450\,^\circ\text{C}$). This conclusion was confirmed experimentally [140] and the Sb$_4$ line was found to be the only one in the mass spectrum of Sb molecules evaporated from a Si surface up to 820K ($\sim 550\,^\circ\text{C}$). For higher temperatures, Sb$_2$ and Sb$_1$ lines appear in the mass spectrum of evaporating molecules (Sb$_4$ flux from a Sb cell was used as a source of Sb molecules). The latter observation also agrees fairly well with the results of the calculations according to (4.1) [53]. Thus, for typical InSb MBE growth conditions, the Sb$_4$ molecules should be considered as equilibrium group-V molecular species. This is in contrast to other calculations [126], where antimony dimer molecules were considered as the equilibrium features.

According to the results obtained in [53], we can write the reaction for InSb growth (evaporation) as:

$$\text{In(g)} + \frac{1}{4}\text{Sb}_4(\text{g}) \Leftrightarrow \text{InSb(s)} \tag{4.2}$$

and

$$p_{\text{In}} p_{\text{Sb}_4}^{1/4} = (K_{\text{InSb}})^{-1} = 7.68 \times 10^8 \exp\left(\frac{-3.31\,\text{eV}}{kT}\right), \tag{4.3}$$

where K_{InSb} is derived according to Table 4.1. The equilibrium constant for the minority antimony dimeric molecules is, according to Table 4.1:

$$p_{\text{In}} p_{\text{Sb}_2}^{1/2} = \left(K_{\text{InSb}}^{\text{Sb}_2}\right)^{-1} = 1.33 \times 10^{11} \exp\left(\frac{-4.1\,\text{eV}}{kT}\right). \tag{4.4}$$

This latter agrees fairly well with the value given by [104] (see Table 2.1).

The minimum antimony pressure in the incident beam necessary to avoid the formation of In droplet on the surface coincides with that corresponding to the transition from (1×3)Sb- to $c(8 \times 2)$In-rich surface reconstruction, revealed in RHEED studies, and that pressure can be written as

$$p_{\text{Sb}_4,\min}^0 = \frac{1}{4}\sqrt{\frac{m_{\text{Sb}_4}}{m_{\text{In}}}} \left(p_{\text{In}}^0 - p_{\text{In}}^{\text{In}_{\text{L}}}\right) + p_{\text{Sb}_4}^{\text{In}_{\text{L}}}, \tag{4.5}$$

where m_{Sb_4} and m_{In} are the respective molecular masses, and $p_{\text{In}}^{\text{In}_{\text{L}}}$ and $p_{\text{Sb}_4}^{\text{In}_{\text{L}}}$ are the respective In and Sb$_4$ equilibrium partial pressures over the In–InSb liquidus of the InSb phase diagram.

Furthermore, the expansion for $p_{\text{In}}^{\text{In}_{\text{L}}}$ is as follows:

$$p_{\text{In}}^{\text{In}_{\text{L}}}(\text{InSb}) = p_{\text{In}}^{\text{L}}[\text{In}_{\text{L}}]\gamma_{\text{In}}, [\text{In}_{\text{L}}] + [\text{Sb}_{\text{L}}] = 1, \tag{4.6}$$

where γ_{In} and $[\text{In}_{\text{L}}]$ are, respectively, the In activity coefficient and concentration in the liquid phase, and p_{In}^{L} is the indium equilibrium pressure over the pure indium melt. Using the experimental expression for the Sb concentration in the liquid phase ($[\mathrm{Sb_L}]$) according to [93], we can obtain for $T_\mathrm{S} = 320 - 480\,°\mathrm{C}$:

$$[\mathrm{Sb_L}] = 47 \exp\left(\frac{-0.33\,\mathrm{eV}}{kT}\right) \tag{4.7}$$

and

$$\gamma_\mathrm{In} = \exp\left(\frac{\alpha(T)\,[\mathrm{Sb_L}]}{kT}\right), \tag{4.8}$$

where $\alpha(T)$ is $3.4T$ to $12T$ (cal/mol) or $1.5\times 10^{-4}T$ to $5.2\times 10^{-4}T$ (eV) and is the interaction parameter in the liquid phase [93]. As the concentration $[\mathrm{Sb_L}]$ increases noticeably at $T_\mathrm{S} > 350\,°\mathrm{C}$ such that γ_In cannot be neglected, we can obtain from (4.6), (4.7) and (4.8), using the parameters from Table 4.1:

$$p_\mathrm{In}^\mathrm{In_L} = 1.03 \times 10^5 \exp\left(-\frac{2.4\,\mathrm{eV}}{kT}\right) \gamma_\mathrm{In}\,(1 - [\mathrm{Sb_L}])\,, \tag{4.9}$$

and, from (4.6):

$$p_\mathrm{Sb_4}^\mathrm{In_L} = 3.09 \times 10^{15} \exp\left(-\frac{3.64\,\mathrm{eV}}{kT}\right) \gamma_\mathrm{In}^{-4}\,(1 - [\mathrm{Sb_L}])^{-4}\,. \tag{4.10}$$

The respective equation for the $\mathrm{Sb_2}$ equilibrium pressure can be derived from (4.4) and (4.9) as:

$$p_\mathrm{Sb_2}^\mathrm{In_L} = 1.66 \times 10^{12} \exp\left(-\frac{3.4\,\mathrm{eV}}{kT}\right) \gamma_\mathrm{In}^{-2}\,(1 - [\mathrm{Sb_L}])^{-2}\,. \tag{4.11}$$

The expression for the $\mathrm{Sb_4}$ pressure over the pure Sb at which Sb(s) starts to precipitate on the InSb growth surface is:

$$p_\mathrm{Sb_4}^\mathrm{Sb_S} = 3.26 \times 10^7 \exp\left(-\frac{2.0\,\mathrm{eV}}{kT}\right). \tag{4.12}$$

For the heteroepitaxial growth of InSb on GaAs, assuming that no strain relaxation occurs on the initial stage of growth, we need to take into consideration the additional strain-induced Gibbs free energy (ΔG_str) (see [53, 117, 130]):

$$\Delta H_\mathrm{str} = 2G \left[\frac{(1+\nu)}{(1-\nu)}\right] V_m \left[\frac{a - a_0}{a_0}\right]^2, \tag{4.13}$$

where: $\nu = C_{12}/(C_{11} + C_{12})$ is the Poisson ratio; V_m is the substrate molar volume (28.6 cal^3/mol [130]); $G = 1/2 C_{44}$ (for the (100) plane); C_{11}, C_{12} and C_{44} are the InSb elastic coefficients ($1.57 \times 10^4, 0.85 \times 10^4, 0.71 \times 10^4$ cal/cm^3, respectively [101]); a is the InSb lattice constant (6.47937×10^{-8} cm). The calculation according to (4.13) with the above parameters gives $\Delta G_{\text{str,max}} = 0.39\,\text{eV}(\Delta a/a = 14.6\%)$. Since $p_{\text{In}}^{\text{In}_\text{L}}$, the pressure over the In–InSb liquidus, is not affected by the presence of strain (given that liquid is incompressible), $p_{\text{Sb}_4}^{\text{In}_\text{L}}$ is the only variable parameter in (4.3) that can change as a result of a ΔG_{str} change. Because during pseudomorphic growth the crystalline order is not assumed to change significantly (although the transition from cubic to tetragonal symmetry occurs), the change in the reaction entropy can be neglected, i.e. $\Delta S_{\text{str}} \approx 0$. Hence, introduction of the additional strain-induced Gibbs free energy, $\Delta G_{\text{str}} = 0.39\,\text{eV}$, results in the following equation for the Sb_4 equilibrium pressure over the In–InSb liquidus:

$$\left(p_{\text{Sb}_4}^{\text{In}_\text{L}}\right)_{\text{str}} = 3.09 \times 10^{15} \exp\left(-\frac{2.08\,\text{eV}}{kT}\right) \gamma_{\text{In}}^{-4} \left(1 - [\text{Sb}_\text{L}]\right)^{-4}. \qquad (4.14)$$

In Fig. 4.1 we show the calculated temperature dependencies according to: (4.9), (4.10), (4.14) (*solid lines*), (4.12) (*broken line*), (4.11) (*dotted line*), and (4.5) for different growth rates (*dashed lines 1–3*). The values of $p_{\text{Sb}_4,\text{exc}}$ estimated from the experimental data of [134, 140] are presented as dashed lines 4 and 5 respectively. The calculated dependencies for the unstrained case are shown as well, because the InSb growth on the GaAs substrate proceeds like a homo-epitaxial growth after the lattice mismatch is accommodated by a high density of dislocations.

At the same time, on the assumption of pseudomorphic growth of InSb on GaAs, the antimony partial pressure over the In–InSb liquidus increases dramatically to values that are practically unreachable under MBE conditions (see Fig. 4.1, $(p_{\text{Sb}_4}^{\text{In}_\text{L}})_{\text{str}}$. In this case, the incident antimony flux cannot suppress a re-evaporation of antimony species from the surface in the T_S range of interest, resulting in a depletion of the surface with Sb molecules. Hence, after the formation of a first monolayer (which cannot be treated as an InSb solid phase because the atoms are in a very different environment as compared with the bulk), the formation of subsequent monolayers of an InSb solid phase turns out to be impossible until the strain is relaxed.

This relaxation process, according to the above considerations, should occur, first, by a formation of this In-rich quasiliquid phase heavily oversaturated with antimony, as compared with the homo-epitaxial growth in In-rich conditions. With further In atoms arriving at the surface, this quasiliquid phase gradually increases its volume until the critical overstaturation of the liquid phase with Sb forms three-dimensional elastically relaxed InSb islands that are energetically favorable. (The conditions concerning the critical size of the islands are considered in the forthcoming section.)

After formation, the islands exhibit ripening, at the expence of the quasiliquid phase as it is oversaturated with respect to unstrained InSb, and finally an array of either microscopic or macroscopic 3D islands is formed on the surface. Formation of 3D islands seems to be unavoidable unless the

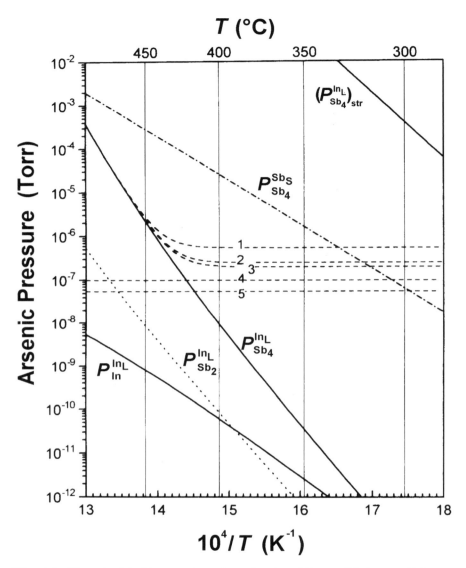

Fig. 4.1. The calculated temperature dependencies of the equilibrium partial pressures of: In and Sb_4 along the In–InSb liquidus ($p_{In}^{In_L}$ and $p_{Sb_4}^{In_L}$, respectively, *solid lines*); Sb_4 along the In–InSb in the presense of strain [$(p_{Sb_4}^{In_L})_{str}$, *solid line*]; Sb_2 along the In–InSb liquidus ($p_{Sb_2}^{In_L}$, *dotted line*); Sb_4 along the Sb–InSb solidus ($p_{Sb_4}^{Sb_S}$, *broken line*); the minimum possible Sb_4 beam pressure ($p_{Sb_4,\min}$, *dashed lines*) at different growth rates [(1) 1 µm/h, (2) 0.55 µm/h, (3) 0.5 µm/h]. The dashed horizontal lines (4) and (5) correspond to the Sb_4 excess equilibrium pressures ($p_{Sb_4,exc}$)

excess Sb pressure is comparable with that corresponding to $p_{\text{Sb}_4}^{\text{In}_L}$, and even in this case the partly relaxed islands will not result in a disappearence of the quasiliquid phase because the latter are not oversaturated with respect to unstrained InSb. Fast growth of the islands at the expense of the quasiliquid phase is, thus, not the case. The formation of 3D islands can be expected to be much less effective in this case. The effective thickness of the quasiliquid will be much greater unless mesoscopic islands of liquid are formed. The growth kinetic studies are in an agreement with the current scenario [142].

We can see from Fig. 4.1 that the indium evaporation pressure ($p_{\text{In}}^{\text{In}_L}$) can be neglected up to an InSb melting point ($\sim 525\,^\circ$C) at commonly used V_g values ($\geq 0.1\,\mu$m/h). Since $p_{\text{In}}^{\text{In}_L}$ is the maximum possible In equilibrium pressure in the system, as the indium equilibrium pressure is always lower over the field of solidus, the InSb growth rate does not decrease with substrate temperature in the T_S range of interest (300–500 $^\circ$C). At the same time, the drastic increase in the Sb$_4$ equilibrium pressure over the liquidus, $p_{\text{Sb}_4}^{\text{In}_L}$, at $T_S > 420\,^\circ$C results in a subsequent increase in $p_{\text{Sb}_4,\text{min}}$, which is needed for the growth under the Sb-stabilized conditions (Fig. 4.1, curves 1, 2, 3). In practice, this results in the necessity of a larger Sb$_4$/In flux ratio as compared with growth at low T_S.

A very important difference of the considered system from any other III–V compounds is a very low antimony equilibrium pressure over the Sb–InSb solidus $\left(p_{\text{Sb}_4}^{\text{Sb}_S}\right)$ [see (4.12) and Fig. 4.1, *broken line*] at the temperatures typically used for growth. This pressure should be compared with the excess antimony beam pressure ($p_{\text{Sb}_4,\text{exc}}$) during growth, which can be determined in the following way:

$$p_{\text{Sb}_4,\text{exc}} \approx p_{\text{Sb}_4}^0 - \frac{1}{4}\sqrt{\frac{m_{\text{Sb}_4}}{m_{\text{In}}}}\left(p_{\text{In}}^0 - p_{\text{In}}^{\text{In}_L}\right), \tag{4.15}$$

where $p_{\text{Sb}_4,\text{exc}}^0$ and p_{In}^0 are the Sb$_4$ and In incident pressures respectively. If $p_{\text{Sb}_4,\text{exc}}$ becomes higher than $p_{\text{Sb}_4}^{\text{Sb}_S}$ during growth, then the precipitates of the antimony solid phase (Sb(s)) appear on the surface, resulting in the disturbance of the layer morphology.

This phenomenon has been experimentally observed [140] as a remarkable decrease of the Sb$_4$ desorption flux from the Si substrate surface at $T_S \approx 270\,^\circ$C and at incident Sb$_4$ flux intensity $J_{\text{Sb}_4} = 6.4 \times 10^{12}$ mls cm^{-2}s^{-1} ($J_{\text{In}} = 0$), which corresponds to $p_{\text{Sb}_4}^0 = p_{\text{Sb}_4,\text{exc}}^0 = 5.4 \times 10^{-8}$ Torr (Fig. 4.1, *horizontal line 5*). The point of interception of line 5 with the $p_{\text{Sb}_4}^{\text{Sb}_S}$ temperature dependence gives $T_S = 295\,^\circ$C, which is in good agreement with $T_S = 270\,^\circ$C obtained experimentally [140]. It was also found [134] that at low substrate temperatures (T_S less than $\sim 270\,^\circ$C) single crystalline InSb films could not be obtained at the given growth rate of 0.5 μm/h (see $p_{\text{Sb}_4}^0$ in Fig. 4.1, curve 3) and an Sb$_4$/In flux ratio of 1.4, because a large concentration of Sb precipitates appeared on the surface. The excess Sb$_4$ pressure, recalculated for these growth parameters, equals $p_{\text{Sb}_4,\text{exc}} = 9.6 \times 10^{-8}\,\text{Torr}$ (see line 4, Fig. 4.1) and it intersects the $p_{\text{Sb}_4}^{\text{Sbs}}$ dependence at $T_S = 305\,^\circ\text{C}$.

The results of these thermodynamic predictions agree fairly well with experimental results, showing that the single crystalline InSb films with mirror-like surface morphology could be obtained at low $T_S \approx 350 - 380\,^\circ\text{C}$ only in a narrow range of Sb_4/In flux ratios (particularly at high growth rates $> 1\,\mu\text{m/h}$ [17, 92] while at higher substrate temperatures $T_S = 450\,^\circ\text{C}$ but $v_G = 0.5\,\mu\text{m/h}$ it is possible to see growth occur in a wide range of flux ratios (1–8) [134] without any degradation of morphology. As an example, it can be noted that to provide an excess antimony pressure $p_{\text{Sb}_4,\text{exc}} > p_{\text{Sb}_4}^{\text{Sbs}}$ at $T_S = 350\,^\circ\text{C}$ and at $v_G = 0.5\,\mu\text{m/h}$, it is necessary to have the Sb_4/In flux ratio higher than 5 (provided that $\text{Sb}_4/\text{In} = 1$ corresponds to the transition between the $(1 \times 3)\text{Sb}$ and the $c(8 \times 2)\text{In}$ surface reconstruction), whereas at $T_S = 420\,^\circ\text{C}$ the formation of antimony precipitates on the surface can start only with an Sb_4/In flux ratio as high as 50.

As can be seen, the results of the calculation appear to be in a very satisfactory agreement with the experimental data, indicating that the thermodynamic approach is very useful in describing surface effects during molecular beam epitaxy.

4.2 Growth of Lattice-Matched Solid Solution Formed From Lattice-Mismatched Binaries

The growth of alloys composed of binaries having significant lattice mismatch can result in the spontaneous formation of compositionally modulated structures [48, 49, 61]. The MBE growth proceeds usually at rather low temperatures to provide significant diffusion coefficients in the solid phase, so that the formation of compositionally modulated structures can occur only via the reactions on the surface of the growing film.

In the case of hetero-epitaxial growth of an alloy with an average composition lattice matched to the substrate, compositional domains formed on the surface will nevertheless have a finite lattice mismatch with the substrate. As was considered in the previous section, strain in the epitaxial film dramatically affects the phase equilibria on a crystal surface during growth, and thus can significantly affect the state of the surface. On the other hand, the energetics of the growth itself can stimulate the formation of compositional domains, as has already been shown for the example of lattice-matched AlGaAs growth on a GaAs substrate. The spontaneous formation of compositionally modulated structures during MBE growth have been reported by a number of authors [13, 90, 105, 143]. Formation of In-rich and In-poor regions during the MBE growth of InGaAs on InP(100) substrates [16] and InGaP on a GaAs(100) substrate resulted in a spontaneous formation of dense arrays of quantum wires oriented along the [0$\bar{1}$1] direction [13] have also been reported.

4.2 Growth of Lattice-Matched Solid Solution 67

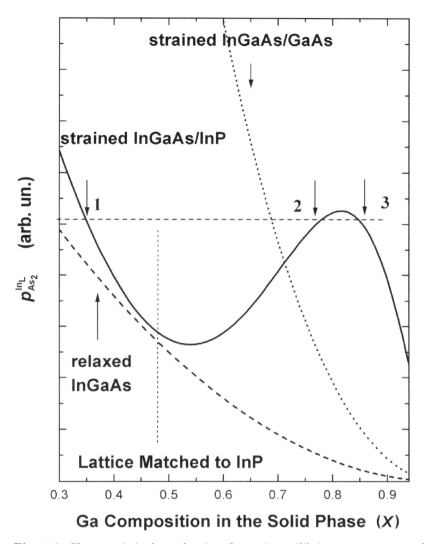

Fig. 4.2. Characteristic dependencies of arsenic equilibrium pressure over the In–Ga liquidus for: relaxed $Ga_xIn_{1-x}As$ (*dashed line*), strained $Ga_xIn_{1-x}As$ on GaAs substrate (*dotted line*), and lattice-matched $Ga_xIn_{1-x}As$ on InP substrate (*solid line*)

In Fig. 4.2 we show qualitatively the dependencies of arsenic equilibrium pressure over the In–Ga liquidus for: relaxed $Ga_xIn_{1-x}As$ (*dashed line*); for strained $Ga_xIn_{1-x}As$ on a GaAs substrate (*dotted line*); and for lattice-matched $Ga_xIn_{1-x}As$ on an InP substrate (*solid line*).

In the case of a thick relaxed $Ga_xIn_{1-x}As$ epilayer, the equilibrium arsenic pressure over the In–Ga liquidus equilibrium with the given x value ($p^L_{As_2}$) increases quadratically with an increase in indium composition [see (2.49) and Fig 4.2 dashed line]. Thus:

$$p^L_{As_2} \approx (1-x)^2 p^{In_L}_{As_2}, \qquad (4.16)$$

where $p^{In_L}_{As_2}$ is the arsenic equilibrium pressure over the In–InAs liquidus. (We neglect here the activity coefficients for simplicity.) This pressure becomes comparable to the excess arsenic pressure ($\sim 10^{-6}$ Torr) typically used in MBE at $\sim 600\,^\circ\text{C}$.

In the case of a strained $Ga_xIn_{1-x}As$ layer, the arsenic pressure over the liquidus (Fig. 4.2, dotted line) is much higher because of the presence of strain in the solid phase (see the preceding section). As has already been discussed for AlGaAs growth, the group-III metal-rich segregation layer appears on the surface if the excess arsenic pressure $p_{As_2,exc}$ becomes smaller than the pressure corresponding to the metal-rich boundary of the more volatile binary compound forming the solid solution. However, for the pseudomorphic $Ga_xIn_{1-x}As$ growth on a GaAs substrate, the equilibrium arsenic pressure over the In-strained InAs liquidus $(p^{In_L}_{As_2})_{str}$ is by several orders of magnitude higher than the $p^{In_L}_{As_2}$ (see the preceding section). Taking into account these considerations, we can see that it is practically impossible to suppress indium segregation at typical MBE growth conditions. There exists, however, an easy way to eliminate the segregated layer: the introduction of growth interruptions at sufficiently high temperatures. The indium atoms forming the wetting layer evaporate, and the layer depletion with indium stops after the first monolayer of the surface-phase $Ga_xIn_{1-x}As$ becomes GaAs-rich. The GaAs is thermally stable up to much higher substrate temperatures, and so the evaporation process automatically stops at this point.

In the case of $Ga_xIn_{1-x}As$ growth on an InP substrate (Fig. 4.2, solid line), the situation changes. A local minimum appears in the relation between $p^{In_L}_{As_2}$ and x at the composition defined by the lattice-matching condition with the substrate. The minimum in $p^{In_L}_{As_2}$ is as a result of two competing processes: the decrease in indium composition on one side, which should decrease the $p^{In_L}_{As_2}$; and, on the other side, the significant strain in the solid phase through its having a significant lattice mismatch with the InP substrate acts in a way to increase $p^{In_L}_{As_2}$. The latter effect is much stronger than the relatively weak dependence of (4.16).

As a result, the $p^{In_L}_{As_2}$ shows clearly nonmonotonic behavior with the indium composition. For a given value of $p_{As_2,exc}$, the indium segregation layer can be in equilibrium with three surface solid-phase monolayer compositions: see points 1, 2 and 3 in Fig. 4.2. The growth surface should then represent of mixture of possible phases (1, 2 or 3), arranged in such a way as to provide for the possibility of the partial relaxation of strains in each of them. This can be achieved if the phases form nanoscale islands on the substrate surface, with a shape determined by the symmetry of the intrinsic stress tensor [109]. For the (100) surface having a large anisotropy along the [010] and [011] directions, a formation of quasiperiodic array of nanoscale stripes along the [0$\bar{1}$1] direction is energetically favorable. For surfaces having a low symmetry (e.g. high-index surfaces), the formation of quantum dots can also be possible.

We can note also that the situation for the growth of relaxed $Ga_xIn_{1-x}As$ film is similar to the case of $Ga_xIn_{1-x}As$ lattice-matched to InP. The composition resulting in the local minimum in $p_{As_2}^{In_L}$ occurs at x, corresponding to average indium composition in the relaxed film.

5. II–VI Materials

A thermodynamic model has been used to describe MBE growth of II–VI compounds [55, 56, 64].

Let us consider II–VI growth. According to [64]:

$$\text{II(g)} + \frac{1}{2}\text{VI}_2(\text{g}) \Leftrightarrow \text{II}-\text{VI(s)}, \tag{5.1}$$

with an equilibrium constant given by

$$p_{\text{II}} p_{\text{VI}_2}^{1/2} = K_{\text{II}-\text{VI}}^{-1}. \tag{5.2}$$

Equilibrium constants for II–VI compounds are listed in Table 5.1. See also Fig. 5.1. According to [64], it is dimeric group-VI element molecules that are in equilibrium with the substrate at typical MBE substrate temperatures and total group-VI element pressures.

Table 5.1. Thermodynamic parameters for the In-Sb system

Compound	$K_{\text{II}-\text{VI}} = p_{\text{II}} p_{\text{VI}_2}^{1/2}$
ZnS	$1.75 \times 10^9 \exp(-3.73/kT)$
ZnSe	$2.76 \times 10^9 \exp(-3.68/kT)$
CdS	$6.90 \times 10^9 \exp(-3.36/kT)$
CdSe	$4.31 \times 10^8 \exp(-3.12/kT)$
ZnTe	$4.31 \times 10^8 \exp(-3.12/kT)$
CdTe	$5.29 \times 10^8 \exp(-2.84/kT)$

The remarkable difference with MBE growth of II–VI compounds is that both group-II and group-VI precipitates cannot form on the substrate surface at typical MBE growth conditions. As follows from Fig. 5.2, only in the case of growth of Mg-containing alloys at substrate temperatures below 250 °C can condensation of Mg be expected in the absence of any group-VI element flux. This unique opportunity gives the possibility of realizing a regime of atomic layer epitaxy widely exploited for II–VI growth.

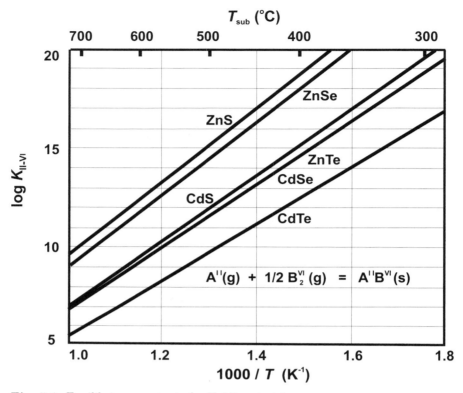

Fig. 5.1. Equilibrium constants for II–VI materials

Another principal difference is related to the much stronger deviation of growth conditions from equilibrium in the case of II–VI growth, as compared to the typical growth regimes for III–V materials. The temperature corresponding to equilibrium conditions for typically used beam pressures for Zn(S,Se) growth is about 560 °C (T_e). In contrast, growth typically occurs at 250–300 °C (T_S). Relative overcooling in this case $(T_e - T_S)/T_S$ is about 0.4; for GaAs MBE growth, it is usually around 0.03.

In spite of this strong deviation from equilibrium, a thermodynamic model has been shown to be in very good agreement with the experimental data for MBE growth of II–VI binaries and of some solid solutions [64]. It should be noted, however, that in the case of Zn(S,Se) and (Zn,Mg)(S,Se) growth, some deviation from theoretical predictions occur [55,56]. This deviation was attributed to kinetic limitations for particular reactions, related to growth conditions strongly out of equilibrium and high vapour pressures of both group-VI and group-II elements. The deviation from the simple mass-rate was accounted for by the introduction of activity coefficients for the gaseous species. It was also demonstrated, using thermodynamic considerations in

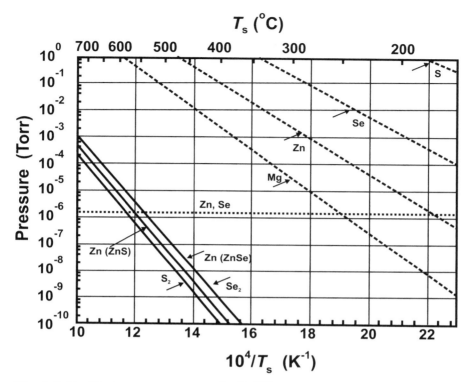

Fig. 5.2. Equilibrium vapor pressures for II–VI binaries and group-II and group-VI elements

the calculations [78], that II–VI solid solutions can be unstable with respect to spinodal decomposition, in agreement with experimental observations [78, 123]. In the case of spinodal decomposition, the simple relations based on assumptions of uniform solid solutions are not appropriate.

It was also demonstrated that in case of II–VI alloy growth a spontaneous corrugation of the growth surface occurs that is related to formation of quasiperiodic compositionally-modulated structures [123]. Self-organization effects, which significantly alter the local composition and the strain state of the system, certainly affect the growth reactions and may cause deviations from the predictions based on the assumption of a uniform alloy composition and a flat surface for the growing film. These effects require special consideration (see e.g. [9] and references therein) and are beyond the scope of this book.

6. Conclusion

As has been shown throughout this book, a macroscopic thermodynamic model can be directly applied to predict and describe all the most important growth effects in MBE. It is particularly important when it is necessary to start working with a new materials system, and thermodynamic analysis easily provides a suitable range of deposition conditions. Numerous dangerous effects (e.g. ultrafast impurity diffusion, segregation of main elements and impurities) can be foreseen and taken into account. The thermodynamic model gives prerequisites of self-organization effects at crystal surfaces (e.g. the appearance of different phases coexisting on the surface, the general applicability of a thermodynamic framework to the description of such effects).

There has been general success with the theory of thermodynamics being applied to the description of various effects during MBE growth [50–54,62,67] clearly points to its importance for practical work related to the optimization of the quality and morphology of epilayers and to composition control. It gives a proper background for the engineering of semiconductor heterostructures used for research and applications in opto- and microelectronics.

References

1. Addinall R., Murray R., Newman R.C., Wagner J., Parker S.D., Williams S.D., Dropad R., De Olivera A.G., Ferguson I., Stradling R.A. (1991) Semicond. Sci. Technol. **6**, 147
2. Airaksinen V.M., Cheng T.S., and Stanley C.R. (1987) J. Cryst. Growth **84**, 241
3. Alexandre F., Goldstein L., Leroux G., Concour M.C., Thiberge H., Rao E.V.K. (1985) J. Vacuum Sci. Technol. B **3**, 950
4. Alferov Z.I., Ivanov S.V., Kop'ev P.S., Ledentsov N.N., Mel'tser B.Y., Nemenov M.I., Ustinov V.M. (1989) Fiz. Tekh. Poluprovodn. **24**, 152; (1989) Sov. Phys. Semiconductors **24**, 92
5. Amar J.G., Family F. (1995) Phys. Rev. Lett. **74**, 2066
6. Bales G.S., Chrzan D.C. (1995) Phys. Rev. Lett. **74**, 4879
7. Biefeld R.M., Wendt J.R., Kurtz S.R. (1991) J. Crystal Growth **107**, 836
8. Biefeld R.M., Hebner G.A. (1991) J. Crystal Growth **109**, 272
9. Bimberg D., Grundmann M., Ledentsov N.N. (1999) Quantum Dot Heterostructures. Wiley, Chichester. 328 pp
10. Bresse J.F., Papadopulos A.C. (1987) Appl. Phys. Lett. **51**, 183
11. Bressler-Hill V., Varma S., Lorke A., Nosho B.Z., Petroff P.M., Weinberg W.H. (1995) Phys. Rev. Lett. **74**, 3209
12. Casey Jr. H.C., Panish M.B. (1978) Heterostructure Lasers. Academic Press, New York
13. Cheng K.Y., Hsieh K.S., Baillargeon J.N. (1992) Appl. Phys. Lett. **60**, 2892
14. Chiang P.K., Bedair S.M. (1984) J. Electrochem. Soc. **131**, 2422
15. Cho A.Y., Arthur Jr. J.R. (1975) Progr. Solid State Chem. **10**, 157
16. Chou S.T., Cheng K.Y., Chou L.J., Hsieh K.C. (1995) Appl. Phys. Lett. **66**, 2220
17. Chyi J.-I., Kalem S., Kmar N.S., Litton C.W., Morkoç H. (1988) Appl. Phys. Lett. **53**, 1092
18. Chyi J.-I., Biswas Q., Iyer S.V., Kumar N.S., Morkoç H., Bean R., Zanio K., Lee H.-Y., Chen H. (1989) Appl. Phys. Lett. **54**, 1016
19. Clarke S., Vvedensky D.D. (1987) Appl. Phys. Lett. **51**, 340
20. Covington D.W., Meeks E.L. (1979) J. Vac. Sci. Technol. **3**, 531
21. Desimone D., Wood C.E.C., Evans C.A. (1982) J. Appl. Phys. **53**, 4938
22. Duhamel N., Henoc P., Alexandre F., Rao E.V.K. (1981) Appl. Phys. Lett. **39**, 49
23. Dutartre D., Gavand M. (1984) J. Cryst. Growth **66**, 647
24. Enquist P., Wicks G.W., Eastman L.F., Hitzman C. (1985) J. Appl. Phys. **58**, 4310
25. Erickson L.P., Mattord T.J., Palmberg P.W., Fisher R. (1983) Electron. Lett. **19**, 632

26. Evans K.R., Stutz C.E., Lorance D.K., Jones R.L. (1989) J. Vac. Sci. Technol. B **7**, 259
27. Evans K.R., Kaspi R., Ehret J.E., Skowronski M., Jones M. (1995) J. Vac. Sci. Technol. B **13**, 1820
28. Foxon C.T. (1978) Acta Electron. **21**, 139
29. Foxon C.T., Harris J.J., Hilton D., Hewett J., Roberts C. (1989) Semicond. Sci. Technol. **4**, 582
30. Frank F.C., Turnbull D. (1956) Phys. Rev. **104**, 617
31. Fujita S., Bedair S.M., Littlejohn M.A., Hauser J.R. (1980) J. Appl. Phys. **51**, 5438
32. Greene J.E., Barnet S.A., Rockett A., Bajor G. (1985) Appl. Surf. Sci. **22/23**, 520
33. Harmand J.C., Alexandre F., Beerens J. (1987) Rev. Physique Appl. **22**, 821
34. Harris J.J., Ashenford D.E., Foxon C.T., Dobson P.J., Joyce B.A. (1984) Appl. Phys. A **33**, 87
35. Heckingbottom R., Todd C.J., Davies G.J. (1980) J. Electrochem. Soc. **127**, 444
36. Heckingbottom R., Davies G. (1980) J. Cryst. Growth **50**, 644
37. Heckingbottom R., Davies G.J., Prior K.A. (1983) Surf. Sci. **132**, 375
38. Heckingbottom R. (1985) J. Vac. Sci. Technol. B **3**, 572
39. Heckingbottom R. (1985) in: Chang L.L., Ploog K. (eds.) Molecular Beam Epitaxy and Heterostructures. Nijhoff, Dordrecht
40. Horikoshi Y., Kawashima M., Yamaguchi H. (1988) Japan. J. Appl. Phys. **27**, 169
41. Houzay F., Moison J.M., Guille C., Barthe F., Van Rompay M. (1989) J. Cryst. Growth **95**, 35
42. van Hove J.M., Cohen P.I. (1985) Appl. Phys. Lett. **47**, 726
43. Hurle D.T.J. (1979) J. Phys. Chem. Solids **40**, 613
44. Hurle D.T.J. (1979) J. Phys. Chem. Solids **40**, 647
45. Iimura Y., Kawabe M. (1986) Japan. J. Appl. Phys. **25**, L81
46. Ilegems M., Dingle R., Rupp Jr. L.W. (1975) J. Appl. Phys. **46**, 3059
47. Ilegems M. (1977) J. Appl. Phys. **48**, 1278
48. Ipatova I.P., Malyshkin V.G., Shchukin V.A. (1993) J. Appl. Phys. **74**, 7198
49. Ipatova I.P., Malyshkin V.G., Shchukin V.A. (1994) Phil. Mag. B **70**, 557
50. Ivanov S.V., Kop'ev P.S., Ledentsov N.N. (1990) J. Cryst. Growth **104**, 345
51. Ivanov S.V., Kop'ev P.S., Ledentsov N.N. (1990) J. Cryst. Growth **108**, 661
52. Ivanov S.V., Kop'ev P.S., Ledentsov (1991) J. Cryst. Growth **111**, 151
53. Ivanov S.V., Altukhov P.D., Argunova T.S., Bakun A.A., Budza A.A., Chaldyshev V.V., Kovalenko Y.A., Kop'ev P.S., Kutt R.N., B.Ya. Mel'tser, Ruvimov S.S., Sorokin L.M., Shaposhnikov S.V., Ustinov V.M. (1993) Semicond. Sci. Technol. **7**, 347
54. Ivanov S.V., Boudza A.A., Kutt R.N., Ledentsov N.N., Mel'tser B.Y., Ruvimov S.S., Shaposhnikov S.V., Kop'ev P.S. (1995) J. Cryst. Growth **156**, 191
55. Ivanov S.V., Sorokin L.M., Kop'ev P.S., Kim J.R., Jung H.D., Park H.S. (1996) J. Cryst. Growth **159**, 16
56. Ivanov S.V., Sorokin L.M., Krestnikov I., Faleev N., Ber N., Sedova I., Kudryavtsev Y., Kop'ev P. (1998) J. Cryst. Growth **84/185**, 70
57. Juang F.-Y., Nashimoto Y., Bhattacharia P. (1985) J. Appl. Phys. **58**, 1986
58. Kaminski A.Y., Suris R.A. (1996) in: Scheffler M., Zimmermann R. (eds.) Proceedings of the 23rd Int. Conf. on the Physics of Semiconductors, Berlin 1996. Vol. 2, 1337 World Scientific, Singapore

59. Karpov S.Y., Kovalchuk Y.V., Myachin V.E., Pogorelski Y.V. (1993) J. Crys. Growth **129**, 563
60. Kaspi R., Evans K.R. (1995) Appl. Phys. Lett. **67**, 819
61. Khachaturyan A.G. (1983) Theory of Phase Transformations in Solids. Wiley, New York
62. Kop'ev P.S., Ledentsov (1988) Sov. Phys. Semicond. **22**, 1093
63. Kop'ev P.S., Ivanov S.V., Egorov A.Y., Uglov D.Y. (1989) J. Cryst. Growth **96**, 533
64. Koukitu A., Nakai H., Suzuki T., Seki (1987) J. Cryst. Growth **84**, 425
65. Kröger F.A. (1964) The Chemistry of Imperfect Crystals. North-Holland, Amsterdam; Wiley, New York
66. Ledentsov N.N., Ber B.Y., Kop'ev P.S., Ivanov S.V., Mel'tser B.Y., Ustinov V.M. (1985) Sov. Phys: Techn. Phys. **30**, 80
67. Ledentsov N.N. (1987) Ph. D Thesis (in Russian), A.F. Ioffe Physical Technical Institute, Leningrad
68. Ledentsov N.N., Maximov M.V., Kop'ev P.S., Ustinov V.M., Belousov M.V., Mel'tser B.Y., Ivanov S.V., Shchukin V.A., Alferov Z.I., Grundmann M., Bimberg D., Ruvimov S.S., Richter W., Werner P., Gösele U., Heidenreich U., Wang P.D., Sotomayor Torres C.M. (1995) Microelectronics Journal **26**, 871
69. Ledentsov N.N., Grundmann M., Kirstaedter N., Schmidt O., Heitz R., Böhrer J., Bimberg D., Ustinov V.M., Shchukin V.A., Egorov A.Y., Zhukov A.E., Zaitsev S., Kop'ev P.S., Alferov Z.I., Ruvimov S.S., Kosogov A.O., Werner P., Gösele U., Heydenreich J. (1996) Solid State Electronics **40**, 785
70. Lee G.S., Lin Y.F., Bedair S.M., Laiding D. (1985) Appl. Phys. Lett. **47**, 1219
71. Levin J.L., Alexandre F. (1985) Electron. Lett. **21**, 413
72. Liu S., Bönig L., Detch J., Metiu H. (1995) Phys. Rev. Lett. **74**, 4495
73. Lorenz M.R., Binkowski B.B. (1962) J. Electrochem. Soc. **109**, 24
74. Madhukar A. (1983) Surf. Sci. **132**, 344
75. Marchenko V.I. (1981) Sov. Phys. JETP **54**, 605
76. Marchenko V.I. (1981) Sov. Phys. JETP Lett. **33**, 381
77. Massies J., Turco F., Saletes A., Contor J.P. (1987) J. Cryst. Growth **80**, 307
78. Marbeuf A., Druilhe R., Triboulet R., Patriarche G. (1992) J. Cryst. Growth **117**, 10
79. Matsushima Y., Gonda S. (1976) Jpn. J. Appl. Phys. **15**, 2093
80. McLevige W.V., Vaidyanathan K.V., Streetman B.G., Ilegems M., Comas J., Plew L. (1978) Appl. Phys. Lett. **33**, 127
81. Metze G.M., Stall R.A., Wood C.E.C., Eastman L.F. (1980) Appl. Phys. Lett. **37**, 165
82. Metze G.M., Calawa A.R. (1983) Appl. Phys. Lett. **42**, 818
83. Miller R.C., Tsang W.T., Munteanu O. (1982) Appl. Phys. Lett. **41**, 374
84. Miller J.N., Collins D.M., Moll N.J. (1985) Appl. Phys. Lett. **46**, 960
85. Miller D.L., Asbeck P.M. (1985) J. Appl. Phys. **57**, 1816
86. Moison J.M., Guille C., Bensoussan M. (1987) Phys. Rev. Lett. **58**, 2555
87. Moison J.M., Guille C., Houzay F., Barthe F., Van Rompay M. (1989) Phys. Rev. B **40**, 6149
88. Morkoç H., Drummond T.J., Kopp W., Fisher R. (1982) J. Electrochem. Soc. **129**, 824
89. Nakagawa T., Gonda S., Emura S., Shimazu S. (1988) J. Cryst. Growth **87**, 276
90. Nasi L., Salvati G., Mazzer M., Zanotti-Fregonara C. (1996) Appl. Phys. Lett. **68**, 3263
91. Noreika A.J., Greggi J., Takei W.J., Francombe M.H. (1983) Vac. Sci. Technol. A **1**, 558

92. Oh J.E., Bhattacharya P.K., Chen Y.C., Tsukamoto S. (1989) J. Appl. Phys. **66**, 3618
93. Panish M.B., Ilegems M. (1972) in: Reiss H., McCaldin J. (eds.) Progress in Solid State Chemistry. Vol. 7. Pergamon, Oxford
94. Pao Y.C., Hierl T., Cooper T. (1986) J. Appl. Phys. **60**, 201
95. Pao Y.C., Franklin J., Harris J.S. (1989) J. Crystal Growth **95**, 301
96. Pauling L. (1945) The nature of the Chemical Bond Cornell University Press, Ithaca, NY
97. Pehlke E., Moll N., Scheffler M. (1996) in: Scheffler M., Zimmermann R. (eds.) Proceedings of the 23rd Int. Conf. on the Physics of Semiconductors, Berlin, 1996. Vol. 2, 1301. World Scientific, Singapore
98. Phillips J.C. (1981) J. Vac. Sci. Technol. **19**, 545
99. Priester C., Lefebvre I., Allan G., Lannoo M. (1995) in: Lockwood D.J. (ed.) Proceedings of the 22nd Int. Conf. on the Physics of Semiconductors, Vancouver, 1994. Vol. 1, 676. World Scientific, Singapore
100. Rao T.S., Well J.B., Houghton P.C., Baribean J.M., Moore W.T., Noad J.P. (1988) Appl. Phys. Lett. **53**, 51
101. Ryabin B.A., Ostroumov M.A., Swit T.F. (1977) Handbook on Thermodynamic Properties of Substances. Chemistry, Leningrad
102. Schwarz S.A., Venkatesan T., Hwang D.M., Joon H.W., Bhat R., Arakawa Y. (1987) Appl. Phys. Lett. **50**, 281
103. Scott E.G., Andrews D.A., Davies G.J. (1986) J. Vac. Sci. Technol. B **4**, 534
104. Seki H., Koukitu A. (1986) J. Cryst. Growth **78**, 342
105. Seong T.Y., Norman A.G., Ferguson I.T., Booker G.R. (1993) J. Appl. Phys. **73**, 8227
106. Shchukin V.A., Borovkov A.I., Ledentsov N.N., Bimberg D. (1995) Phys. Rev. B **51**, 10104
107. Shchukin V.A., Borovkov A.I., Ledentsov N.N., Kop'ev P.S. (1995) Phys. Rev. B **51**, 17767
108. Shchukin V.A., Ledentsov N.N., Kop'ev P.S., Bimberg D. (1995) Phys. Rev. Lett. **75**, 2968
109. Shchukin V.A. (1996) in: Scheffler M., Zimmermann R. (eds.) Proceedings of the 23rd Int. Conf. on the Physics of Semiconductors, Berlin, 1996. Vol 2, 1261. World Scientific, Singapore
110. Shchukin V.A., Ledentsov N.N., Grundmann M., Kop'ev P.S., Bimberg D. (1996) Surf. Sci. **352-354**, 117
111. Shen J., Chatillon C. (1990) J. Cryst. Growth **106**, 543; ibid 553
112. Sidorov V.G., Vasil'eva L.F., Sabinina I.V., Dvoretsky S.A., Sidorova A.V. (1976) J. Electrochem. Soc. **123**, 699
113. Sirenko A.A., Ruf T., Ledentsov N.N., Egorov A.Y., Kop'ev P.S., Ustinov V.M., Zhukov A.E. (1996) Solid State Commun. **97**, 169
114. Stall R.A., Zilko J., Swaminathan V., Schumacher N., (1985) J. Vac. Sci. Technol. B **3**, 524
115. Strel'chenko S.S., Lebedev V.V. (1984) A^{III}–B^V Compounds (in Russian) Metallurgiya, Moscow
116. Stringfellow G.B., Greene P.E. (1971) J. Electrochem. Soc. **118**, 805
117. Stringfellow G.B. (1972) J. Appl. Phys. **43**, 3455
118. Stringfellow G.B. (1974) J. Cryst. Growth **27**, 21
119. Sun Y.L., Masselink W.T., Fisher R., Klein M.V., Morkoç H., Bajaj K.K. (1984) J. Appl. Phys. **55**, 3554
120. Tersoff J., Tromp R.M. (1993) Phys. Rev. Lett. **70**, 2782
121. Tmar M., Gabriel A., Chatillon C., Ansara I. (1984) J. Cryst. Growth **68**, 557

122. Tmar M., Gabriel A., Chatillon C., Ansara I. , (1984) J. Cryst. Growth **69**, 421
123. Tommiya S., Minatoya R., Itoh S., Nakano K., Morita E., Ishibashi A., Ikeda M. (1996) in: Scheffler M., Zimmermann R. (eds.) Proceedings of the 23rd Int. Conf. on the Physics of Semiconductors, Berlin, 1996. p. 1079 World Scientific, Singapore
124. Tournié E., Trampert A., Ploog K.H. (1994) Europhysics Lett. **25**, 663
125. Trampert A., Tournié E., Ploog K.H. (1994) Phys. Stat. Sol. B **145**, 481
126. Tsao I.J. (1991) J. Cryst. Growth **110**, 595
127. Tsui R.K., Curless J.A., Kramer G.D., Peffley M.S., Wicks G.W. (1986) J. Appl. Phys. **59**, 1508
128. Tuck B. (1985) J. Phys. D (Appl. Phys.) **18**, 557
129. Turco F., Massies J., Contour G.P. (1987) Rev. Phys. Appl. **22**, 827
130. Turco F., Guillaume J.C., Massies J. (1988) J. Crystal Growth **88**, 282
131. Wang P.D., Ledentsov N.N., Sotomayor Torres C.M., Kop'ev P.S., Ustinov V.M. (1995) Appl. Phys. Lett. **66**, 112
132. Weisberg R., Blank J. (1963) Phys. Rev. **131**, 1548
133. Van Vechten J.A. (1975) J. Electrochem. Soc. **122**, 419
134. Williams G.M., Whitehouse C.R., Martin T., Chew N.G., Gullis A.G., Ashley T. (1988) J. Appl. Phys. **63**, 1526
135. Williams G.M., Whitehouse C.R., McConville C.F., Cullis A.G., Ashley T., Courtney S.J., Alliot C.T. (1988) Appl. Phys. Lett. **53**, 1189
136. Wood C.E.C., Joyce B.A. (1978) J. Appl. Phys. **49**, 4854
137. Wood C.E.C., Desimone D., Singer K., Wicks G.W. (1982) J. Appl. Phys. **53**, 4230
138. Wood C.E.C., Stanley C.R., Wicks G.W., Esi M.B. (1983) J. Appl. Phys. **54**, 1868
139. Yano M., Takase T., Kimuta M. (1979) Phys. Status Solidi A **141**, 70
140. Yata M. (1986) Thin Solid Films **137**, 79
141. Yoshida M., Watanabe K. (1985) J. Electrochem. Soc. **132**, 1733
142. Zhang X.X., Staton-Bevan A.E., Pashley D.W., Parker S.D., Droopád D., Williams R.L., Newman R.C. (1990) J. Appl. Phys. **67**, 800
143. Zunger A., Mahajan S. (1994) in: Mahajan S. (vol. ed.) Vol. 3, 1399 in: Moss T.S. (ed.) Handbook on Semiconductors. Elsevier, Amsterdam

Index

3D islands 61

AlAs growth 16
AlGaAs growth 21, 23
AlGaAs growth rate 27, 29
AlGaAs:Sn 46
amphoteric impurities 41

beryllium doping 47
beryllium segregation 48

congruent sublimation 11

diffusion length
– beryllium doping 55
dimeric molecules 10, 13
dopant concentration 36
doping mechanisms 36

equilibrium 8
– binary compounds 15
– ternary compounds 20
equilibrium constants 10
– AlAs growth 16
– binary compounds 15
– GaP growth 19
– II–VI compounds 69
– InAs growth 17
– InP growth 18
– InSb growth 59
equilibrium pressures
– AlGaAs growth 24, 25, 28
– beryllim doping 50
– GaAs growth 14, 23
– GaP growth 20
– in impurity segregation 43
– InAs growth 17
– InP growth 19
– InSb growth 62

free sublimation 10

Ga-rich conditions 15
Ga_2S 40
GaAs growth 5, 15, 31, 33
GaAs phase diagram 9
GaAs:Be 47
GaAs:Ge 42
GaAs:Mn 37
GaAs:Sn 44, 45
GaAs:Zn 39
GaAsP growth 22
GaInAs growth 20
GaInAs strained layers 65, 66
GaP growth 19
Gibbs free energy 60
growth parameters 8
growth processes 5
– quasi-equilibrium nature 7

II–VI compounds 69
– equilibrium constants 69
III–V compounds 4–6, 22
– impurity segregation 46
– interaction parameters 21
impurity concentration 36
impurity segregation 43
impurity segregation layer 46
InAs growth 16
InAs/Si systems 57
InP growth 18
InSb growth 57, 59
InSb/GaAs hetero-epitaxial growth 57
InSb/GaAs systems 57
inverse equilibrium constant 10, 11

kinetic models 6
kinetics 1

lattice mismatch 57, 64
lattice-matched growth 64, 65
lattice-mismatched growth 57
lattice-mismatched substrates 57

liquid phase epitaxy (LPE)   7, 33
LPE (liquid phase epitaxy)   7, 33

manganese doping   37
mass action equation   10, 33
MBE (molecular beam epitaxy)   1
– apparatus   3
molecular beam condition   4
molecular beam epitaxy (MBE)   1
– apparatus   3

partial pressures   10, 12
phase rule   9
point defects   33, 35
point-defect equilibria   33

quantum wires   64
quasi-equilibrium growth   27

reflection high-energy electron diffraction (RHEED)   5
RHEED (reflection high-energy electron diffraction)   5

segregation layer
– beryllium   49, 51, 52, 55, 56
– indium   66
self-organization   22, 57, 71, 73
solid-state interaction parameters   21
sublimation
– congruent   11

– free   10
– suppression   12
sulphur doping   40
suppression of sublimation   12
surface segregation   22, 23
surface solid-phase monolayer   48

temperatures of maximum sublimation   11, 15
– AlAs growth   16
– GaAs growth   11
– GaP growth   20
– III–V compounds   11
– InAs growth   18
– InP growth   18
tetrameric molecules   10, 11
thermodynamic models   7, 8, 13, 38, 41, 70
– macroscopic   73
thermodynamic parameters   58
thermodynamics   1
tin doping   44
total pressure   10

ultrahigh vacuum   4

vapor phase epitaxy (VPE)   7, 33
VPE vapor phase epitaxy   7, 33

zinc doping   39

Springer Tracts in Modern Physics

140 **Exclusive Production of Neutral Vector Mesons at the Electron-Proton Collider HERA**
By J. A. Crittenden 1997. 34 figs. VIII, 108 pages

141 **Disordered Alloys**
Diffusive Scattering and Monte Carlo Simulations
By W. Schweika 1998. 48 figs. X, 126 pages

142 **Phonon Raman Scattering in Semiconductors, Quantum Wells and Superlattices**
Basic Results and Applications
By T. Ruf 1998. 143 figs. VIII, 252 pages

143 **Femtosecond Real-Time Spectroscopy of Small Molecules and Clusters**
By E. Schreiber 1998. 131 figs. XII, 212 pages

144 **New Aspects of Electromagnetic and Acoustic Wave Diffusion**
By POAN Research Group 1998. 31 figs. IX, 117 pages

145 **Handbook of Feynman Path Integrals**
By C. Grosche and F. Steiner 1998. X, 449 pages

146 **Low-Energy Ion Irradiation of Solid Surfaces**
By H. Gnaser 1999. 93 figs. VIII, 293 pages

147 **Dispersion, Complex Analysis and Optical Spectroscopy**
By K.-E. Peiponen, E.M. Vartiainen, and T. Asakura 1999. 46 figs. VIII, 130 pages

148 **X-Ray Scattering from Soft-Matter Thin Films**
Materials Science and Basic Research
By M. Tolan 1999. 98 figs. IX, 197 pages

149 **High-Resolution X-Ray Scattering from Thin Films and Multilayers**
By V. Holý, U. Pietsch, and T. Baumbach 1999. 148 figs. XI, 256 pages

150 **QCD at HERA**
The Hadronic Final State in Deep Inelastic Scattering
By M. Kuhlen 1999. 99 figs. X, 172 pages

151 **Atomic Simulation of Electrooptic and Magnetooptic Oxide Materials**
By H. Donnerberg 1999. 45 figs. VIII, 205 pages

152 **Thermocapillary Convection in Models of Crystal Growth**
By H. Kuhlmann 1999. 101 figs. XVIII, 224 pages

153 **Neutral Kaons**
By R. Belušević 1999. 67 figs. XII, 183 pages

154 **Applied RHEED**
Reflection High-Energy Electron Diffraction During Crystal Growth
By W. Braun 1999. 150 figs. IX, 222 pages

155 **High-Temperature-Superconductor Thin Films at Microwave Frequencies**
By M. Hein 1999. 134 figs. XIV, 395 pages

156 **Growth Processes and Surface Phase Equilibria in Molecular Beam Epitaxy**
By N.N. Ledentsov 1999. 17 figs. VIII, 84 pages

157 **Deposition of Diamond-Like Superhard Materials**
By W. Kulisch 1999. 60 figs. X, 191 pages

Springer and the environment

At Springer we firmly believe that an international science publisher has a special obligation to the environment, and our corporate policies consistently reflect this conviction.

We also expect our business partners – paper mills, printers, packaging manufacturers, etc. – to commit themselves to using materials and production processes that do not harm the environment. The paper in this book is made from low- or no-chlorine pulp and is acid free, in conformance with international standards for paper permanency.